STUDY GUIDE

NANCY J. OBERMEYER

Indiana State University

FOURTH EDITION

DIVERSITY AMID GLOBALIZATION

World Regions, Environment, Development

ROWNTREE   LEWIS   PRICE   WYCKOFF

PEARSON

Prentice Hall

Upper Saddle River, NJ 07458

Publisher, Geosciences: Daniel Kaveney
Editor-in-Chief, Science: Nicole Folchetti
Project Manager: Tim Flem
Assistant Managing Editor, Science: Gina M. Cheselka
Project Manager, Science: Wendy Perez
Supplement Cover Manager: Paul Gourhan
Supplement Cover Designer: Victoria Colotta
Operations Specialist: Amanda A. Smith
Senior Operations Supervisor: Alan Fischer
Marketing Manager: Amy Porubsky
Cover Photo: *A house on stilts in southern Cambodia on the Mekong River.* Credit: MICHAEL S. YAMASHITA / National Geographic Image Collection.

© 2009 Pearson Education, Inc.
Pearson Prentice Hall
Pearson Education, Inc.
Upper Saddle River, NJ 07458

Printed in the United States of America

10 9 8 7 6 5 4 3 2 1

ISBN-13: 978-0-13-601169-9
ISBN-10: 0-13-601169-1

Pearson Education Ltd., *London*
Pearson Education Australia Pty. Ltd., *Sydney*
Pearson Education Singapore, Pte. Ltd.
Pearson Education North Asia Ltd., *Hong Kong*
Pearson Education Canada, Inc., *Toronto*
Pearson Educación de Mexico, S.A. de C.V.
Pearson Education—Japan, *Tokyo*
Pearson Education Malaysia, Pte. Ltd.

Table of Contents

Preface

It is a pleasure to be writing the Study Guide to the fourth edition of *Diversity Amid Globalization.* I teach World Geography spring and fall semesters, and once during the summer just for good measure. As an instructor who uses *Diversity Amid Globalization,* I appreciate the book's excellent coverage of world regions, along with its clarity and the consistency of chapter layout. My students seem to find the book easy to use, too.

Many of my students find the Study Guide to be helpful in identifying the key elements of the course materials based on both the text and the class lectures. The Study Guide itself is designed to highlight the most important points of the textbook. Many of my students bring the Study Guide with them to class, and use it as the basis for their own note-taking. Specifically, they often use a highlighter to identify the points that are most significant, as identified in class lecture. In addition, they write their own notes in the Study Guide to help them remember significant facts and examples.

The Study Guide also includes a practice quiz at the end of each chapter, which you may also find helpful as a study aid. One other feature is the Region Summary, which allows you to record specific details taken from the tables in each chapter as a way to compare countries within a single region, and to compare regions to each other.

I hope you will find this Study Guide to be of help, and I thank you for using it.

Acknowledgments

I want to thank my editor, Amanda Brown, for her continued excellent and patient direction. Thanks to the authors of *Diversity Amid Globalization,* Les Rowntree, Martin Lewis, Marie Price, and William Wyckoff, for writing such a great text and for allowing me to continue on as author of the Instructor's Manual and Study Guide. It is a pleasure to work with such a terrific team.

Nancy Obermeyer
Bloomington, Indiana
March 2008

REGION SUMMARY

REGION NAME _____

TOTAL POPULATION OF THE REGION _____

NUMBER OF COUNTRIES IN THE REGION _____

POPULATION INDICATORS

	Highest (country and value)	**Lowest (country and value)**
Population 2007 (Millions)	_____	_____
Density per sq km	_____	_____
TFR	_____	_____
Percent Urban	_____	_____
Percent<15	_____	_____
Percent>65	_____	_____
Net Migration (Per 1000; 2000-05)	_____	_____

DEVELOPMENT INDICATORS

GNI per capita/PPP 2005	_____	_____
GDP Avg. Annual Growth (2000-2005)	_____	_____
Life Expectancy 2007	_____	_____
% Living on less than $2/day 2006	_____	_____
<5 Mortality 2005	_____	_____
Gender Equity #	_____	_____

10 LARGEST COUNTRIES SUMMARY

TOTAL POPULATION OF THE WORLD   6.6 billion

POPULATION INDICATORS for 10 largest countries

	Highest (country and value)	Lowest (country and value)
Population 2007 (Millions)	China: 1.318	Japan 128
Density per sq km	Bangladesh: 1035	Russia: 8
TFR	Nigeria: 5.9	Russia, Japan: 1.3
Percent Urban	Brazil: 81	Bangladesh: 23
Percent<15	Nigeria: 45	Japan: 14
Percent>65	Nigeria: 3	Japan: 21
Net Migration (Per 1000; 2000-05)	United States: 4	Pakistan: -2.4

DEVELOPMENT INDICATORS

GNI per capita/PPP 2005	United States: $41,950	Nigeria: $1,040
GDP Avg. Annual Growth (2000-2005)	China: 9.6%	Japan: 1.1%
Life Expectancy 2007	Japan: 82	Nigeria: 47
% Living on less than $2/day 2006	Nigeria: 92%	United States, Japan: 0%
<5 Mortality 2005	Nigeria: 194 per 1000	Japan: 4 per 1000
Gender Equity #	Russia: 110	Pakistan: 75

Chapter 1
DIVERSITY AMID GLOBALIZATION

LEARNING OBJECTIVES

- A key goal of this book is to explain "...how geographic diversity comes in conflict with globalization... Because of the importance of this theme, diversity and globalization should be examined as inseparable --- often in conflict, yet at other times complementary." (Rowntree et al., p. 11).

- This first chapter is one of the most important of the book, because it introduces the themes and concepts that the authors use throughout the text and lays the foundation for understanding the remaining chapters. The themes and concepts developed in this chapter are used in each succeeding chapter as a framework for learning about the regions of the world. In addition to helping the student to assimilate and understand the material for each region, this framework will also pave the way for the student to compare and contrast regions throughout the remainder of the text. **It is critical that the student thoroughly understands the material presented in this chapter.**

- This chapter introduces and explains a number of foundation concepts, including the concept of geography itself.

- This chapter also introduces the fundamental concepts of diversity and globalization, and explains the dynamic tension between them.

- As well, this chapter introduces the themes that will appear in each chapter associated with a region of the world: environmental geography, population and settlement, as well as cultural, political, and economic geography.

- When the student has completed this chapter, he or she should have a solid foundation for assimilating the knowledge about regions that will be presented later in the book. Among the specific concepts and models that the student should understand are the following:
  - The dynamic tension between diversity and globalization and its relevance to the study of geography
  - Region as a geographic concept
  - Culture as an important component of geography
  - Vital statistics (birth, death, and growth rates)
  - The demographic transition
  - Migration as a dynamic force on the globe
  - Urbanization
  - Themes in political geography, including nation-state, centrifugal and centripetal forces, boundaries and frontiers, colonialism, and international and supranational organizations
  - Economic development indicators
  - Social development indicators

CHAPTER OUTLINE

I. Globalization and Diversity: A Geography for the 21st Century

A. Globalization: the increasing interconnectedness of people and places through converging processes of economic, political, and cultural change

B. Converging Currents of Globalization

1. Global communication

2. Global transportation
3. Powerful transnational corporations and financial institutions
4. International free trade agreements
5. Market economies and privatization
6. Homogeneous global consumer culture
7. Economic inequity, disparities
8. International managers, workers

C. Globalization and Cultural Change
1. Spread of western consumer culture challenges local culture; Hybridization: melding of popular and local culture
2. Globalization and Geopolitics: economic and political activity have become intertwined
3. Environmental concerns: economic globalization can harm environments but can also inspire treaties
4. Social dimensions: migration, crime

D. Advocates and Critics of Globalization: The Proglobalization Stance
1. Globalization is the logical expression of modern international capitalism
2. New wealth with trickle down from rich to poor (countries and individuals)
3. Globalization will spread benefits of new ideas and technologies
4. Countries can produce the goods for which they are best suited
5. The need to attract capital from abroad will force countries to adopt sound economic policies

6. World's poor countries will catch up to the rich ones

E. Critics of Globalization
1. Globalization is not "natural"; it's an economic policy promoted by core countries that results in inequities between "haves" and "have nots"
2. Globalization promotes free-market, export-oriented economies at the expense of localized, sustainable economies
3. The "free market" economic model promoted for developing countries is not the one that Western industrial countries used for their own development; instead, government intervention shaped investment in core countries
4. Globalization encourages "bubble" economies

F. A Middle Position?
1. Economic globalization is probably unavoidable
2. Even anti-globalization movement is aided by the Internet – a global technology
3. Countries can "make globalization work" by investing in education and maintaining social cohesion
4. Globalization can be managed

G. Diversity in a Globalizing World
1. Will globalization bring a culturally homogeneous, bland world?
2. There is still much that is different and distinctive about regions of Earth
3. Ethnic and cultural differences foster separatist political movements
4. Uneven economic development

5. Diversity and globalization are inseparable and synergistic

II. Geography Matters: Environments, Regions, Landscapes

A. What Is Geography?
1. Geography describes the Earth and explains patterns on its surfaces
B. Geography Has 2 Complementary Thematic Activities
1. Physical geography examines climates, landforms, soils, vegetation, hydrology (land)
2. Human geography explores social, economic, and political factors; demography, migration, culture (people)
3. Geography explores both topics (e.g., climatology or cultural geography) and regions (e.g., North America or Southeast Asia)
C. Human-Environment Interaction
1. Complex reciprocal interactive relationship between humans and their environment; each influences the other
2. Regional geography compares the ways that various peoples and cultures interact with their unique environments
D. Areal Differentiation and Integration
1. *Areal:* pertains to area
2. *Areal differentiation:* explains differences that mark off one part of the world from another and set it off as a distinct region or place
3. *Areal integration:* the study of how areas interact with each other
E. Regions
1. Geography makes sense of the world by compressing and synthesizing vast amounts of information into spatial categories (regions) that share similar traits
2. Few regions are homogeneous throughout, and differences increase with distance from the center
3. Regional borders (boundaries) are subjective and artificial
F. Space into Place: The Cultural Landscape: cultures vary, but the visible, material expression of human settlement, past and present, shows up on the environment
G. Scales (Global to Local): the size or geographical extent of the area being studied (local, regional, or global)
H. Themes and Diversity Amid Globalization: Environment, population, culture, political geography, economic development

III. Population and Settlement: People on the Land

A. The Earth has more than 6.6 billion people; 129 million are added each year (about 14,000/hour); 90 percent of population growth occurs in developing areas
B. Elements of Population Study
1. Population growth rates vary greatly among the world's regions
2. Regions have vastly different approaches to family planning (to increase or decrease population)
3. Migration is very important; the greatest migration (in terms of number of migrants) in human history is NOW
C. Population Growth and Change
1. *Rate of natural increase (RNI)* – world average is 1.2 percent/yr:

the annual growth rate for a country or region, expressed as a percentage increase or decrease, equals the number of births minus the number of deaths; excluding migration

2. *Total fertility rate (TFR):* the average number of children borne by a statistically average woman; world TFR is 2.7, but ranges from 1.5 (Europe) to 5.0 (Africa); if TFR=2.1, population is stable; if TFR>2.1, population is growing; if TFR<2.1, population will decline

3. *Crude birth rate (CBR):* the gross number of births divided by the total population, giving a figure per 1,000 of the population; world CBR is 21 per 1,000

4. *Crude death rate (CDR):* the gross (total) number of deaths divided by the total population, producing a figure per 1,000 of the population; world CDR is 9 per 1,000

5. Young and Old Populations: Percentage of population under age 15: global average is 30 percent, low is 17 percent (Europe); high of 42 percent (Africa); high number indicates great potential for future growth; Percentage of population over age 65: identifies need for health care and other social services; Population pyramids graphically represent age and gender distribution of a population

6. *Life expectancy:* average number of years a person is expected to live; affected by many factors

D. The Demographic Transition (Figure 1.22)
   1. Change from high birth and death rates to low CBR and CDR (transition means change)
   2. Four stages
      Stage 1: High birth rate, high death rate; very slow growth, low RNI
      Stage 2: Death rate falls dramatically, birth rate remains high; high RNI
      Stage 3: Death rate remains low, birth rate drops; RNI slows
      Stage 4: Death rate low, birth rate low; low RNI (as in stage 1)
   3. Highest RNI during stage 2
E. Migration Patterns
   1. *Pull forces:* good conditions at a destination that cause people to immigrate to an area (economic opportunity, freedom, good climate)
   2. *Push forces:* negative conditions in a person's homeland that cause people to emigrate from (leave) an area (religious or political oppression, war, unemployment)
   3. *Net Migration:* Statistic that shows whether more people are entering or leaving a country
   4. *Rural-to-Urban migration:* Loss of jobs in agriculture is causing rural-to-urban migration worldwide
F. Settlement Geography
   1. Settlement can be clustered or dispersed, and is influenced by economic, political, cultural, and environmental patterns

2. *Population Density:* number of people per unit area of land (sq mi, sq km, etc.)
G. An Urban World
   Urbanized population: percentage of a country's people who live in cities
   Europe, Japan, Australia, United States are about 75 percent urbanized; developing countries about 50 percent or lower; Sub-Saharan Africa about 30 percent urban
H. Conceptualizing the City
   1. *Urban primacy:* describes a city that is disproportionately large and/or dominates economic, political, cultural activities of a country (also called *primate city*)
   2. *Urban structure:* the distribution and pattern of land use within a city (central business district, retail, industry, housing, green space, etc.)
   3. *Urban form:* physical arrangement of buildings, streets, parks, architecture that gives each city its unique sense of place
   4. *Over-urbanization:* occurs when the urban population grows more quickly than services to support the people (e.g., jobs, housing, transportation, sewer, water, electrical lines)
   5. *Squatter settlements:* illegal developments of makeshift housing on land neither owned nor rented by the settlers (more common in developing countries)

III. **Cultural Coherence and Diversity: The Geography of Tradition and Change**
   A. Culture in a Globalizing World (types of culture)
      1. *Culture* is learned (not innate), is shared behavior (not individual); it is held in common by a group of people, empowering them with a "way of life"; culture includes both abstract and material dimensions
      2. *Abstract culture* includes speech and religion
      3. *Material culture* includes technology and housing
      4. Culture changes over time
   B. When Cultures Collide
      1. *Cultural imperialism:* active promotion of one cultural system over another
      2. *Cultural nationalism:* the process of protecting and defending a certain cultural system against diluting or offensive cultural expressions while at the same time actively promoting indigenous culture
      3. *Cultural syncretization* or *hybridization:* blending of cultures to form a new type of culture
   C. Language and Culture in Global Context
      1. Language and culture are intertwined
      2. *Dialect:* a distinctiveness associated with a specific language (e.g., American and British English)
      3. *Lingua franca:* a third language that is adopted by people from different cultural

groups who cannot speak each other's language (e.g., Swahili the lingua franca of Africa)

    D. A Geography of World Religions
      1. *Universalizing religion:* attempts to appeal to all peoples regardless of place or culture (e.g., Christianity, Islam, Buddhism, Mormonism)
      2. *Ethnic religion:* identified closely with a specific ethnic group; faiths that usually do not seek new converts (e.g., Judaism, Hinduism)
      3. *Secularization:* exists when people consider themselves to be either non-religious or outright atheistic

IV. Geopolitical Framework: Fragmentation and Unity

    A. *Geopolitics:* describes and explains the close link between geography and political activity, and focuses on the interaction between power, territory, and space, at all scales
      1. End of Cold War brought hope of stabilization ("New World Order")
      2. Rise in ethnic tensions has occurred instead
    B. Global Terrorism
      1. Attack on World Trade Center on September 11, 2001
      2. Reminder of interconnections between political activity, cultural identity, economic linkages in the world
      3. A product and expression of globalization
      4. *Asymmetrical warfare:* the difference between a superpower's military technology and strategy and

lower level technology and guerilla tactics used by groups like Al Qaeda and the Taliban

    C. *Nation-states:* relatively homogenous cultural group with its own political territory (relatively rare; Japan is an example) Many ethnic groups do not control their land (Palestinians, Kurds, Catalans, Basques)
    D. Centrifugal (disunifying) and Centripetal (unifying) Forces
      1. Centrifugal forces (disunify): linguistic minority status, ethnic separatism, territorial autonomy, disparities in income and well-being
      2. Centripetal forces (unify) shared sense of history, need for national security, overarching economic structure, advantages from larger unified political structure to build and maintain infrastructure (highways, airports, schools)
    E. Boundaries and Frontiers
      1. *Ethnographic boundaries:* borders that follow cultural traits such as boundaries or religion (e.g., Bosnia)
      2. *Geometric boundaries:* perfectly straight lines, drawn without regard for physical or cultural features that usually follow a parallel of latitude or meridian of longitude (e.g., United States-Canada)
    F. Colonialism and Decolonialization
      1. European colonial power has been an important influence on today's world.
      2. *Colonialism:* formal establishment of governmental rule over a foreign population
      3. *Decolonialization:* process of a colony gaining (or regaining)

control over its territory and establishing a separate, independent government

V. Economic and Social Development: The Geography of Wealth and Poverty

A. More- and Less-Developed Countries
1. Core-periphery model of development puts G-8 countries of the northern hemisphere (United States, Canada, France, England, Germany, Italy, Japan, Russia) at the core and all other countries (found in the southern hemisphere) in the periphery
2. *"Third World" countries:* refers to the developing world; the term comes from the Cold War era, describes countries that were not allied with democratic, mainly capitalistic (First World) or communist (Second World) superpowers.
3. Today, we talk about "More Developed Countries" (MDC) or "Less Developed Countries" (LDC)

B. Indicators of Economic Development
1. *Development:* qualitative and quantitative measures indicating structural changes with accompanying changes in the use of labor, capital, and technology; example: change from agricultural to industrial base.
2. *Growth:* increase in the size of a system; e.g., the agricultural or industrial output (product) of a country may grow

3. When a system grows, it gets bigger; when it develops, it gets better.

C. Measuring Economic Wealth: Gross Domestic Product and Income
1. *Gross national income* (GNI): Includes the value of all final goods and services produced within a country's borders *plus* the net income from abroad; this omits non-market economic activity (bartering, household work) and does not consider the degradation or depletion of natural resources that may constrain future economic growth (e.g., clear-cutting forests)
2. *GNI per capita*: divide GNI by the country's population; this adjusts for varying population size
3. *Purchasing power parity* (PPP): gives a comparable figure for a standard "market basket" of goods and services purchased with a local currency to adjust for currency inflation and local cost of living
4. *Economic growth rate:* annual rate of expansion for GDP

D. Indicators of Social Development: measure conditions and quality of human life
1. Life expectancy: average length of life expected at birth for a hypothetical male or female based on national death statistics; average world life expectancy is 66 (range from 46–82)
2. *Percent of population living on less than $2/day:* United Nations measure of extreme poverty

7

3. *Under-age-5 mortality:* number of children who die per 1000 people in the population; influenced by health care, sanitation, availability of food
4. *Gender equity:* ratio of male to females in primary and secondary schools; if ratio > 100, more females than males are in school; if ratio<100, more males than females are in school

VI. Conclusion
A. Discussion of each region to follow includes five themes
   1. Environmental Geography
   2. Population and Settlement
   3. Cultural Coherence and Diversity: A Geographical Mosaic
   4. Geopolitical Framework: Patterns of Dominance and Division
   5. Economic and Social Development Geographies
B. Concepts within these themes
   1. Many environmental issues are global
   2. Many developing regions face rapid population growth; migration to new centers of economic activity; rapid pace of urbanization; concern about whether cities can keep up with demands for jobs, housing, transportation, etc.
   3. Tension between forces of cultural homogenization results from globalization and the counter-currents of small-scale cultural and ethnic identity.
   4. In many regions, geopolitical issues are dominated by ethnic strife and territorial disputes, border tensions with neighbors
   5. Economic and social development are dominated by the increasing disparity between the rich and poor, on both the individual and the State levels.

PRACTICE MULTIPLE CHOICE QUIZ

1. What is the major component of globalization?
   a. Economic reorganization of the world
   b. Global transportation
   c. International athletic competition
   d. Global warming
   e. World communication

2. Which of the following is an example of the globalization of non-material culture?
   a. Eating Thai food in London, England
   b. A Turkish shopkeeper speaking English to an American tourist in Istanbul
   c. A Honda automobile assembly plant in Greensburg, Indiana
   d. A McDonald's in Moscow, Russia
   e. All of the above

3. All of the following statements about globalization are true, EXCEPT...
   a. Cultural globalization is a one-way flow that spreads from the United States to the rest of the world
   b. Globalization has profound geopolitical implications
   c. Globalization is aggravating worldwide environmental problems
   d. Resources previously used only by small local groups are now viewed as global commodities
   e. There is a significant criminal element to contemporary globalization

4. Which of the following people would be LEAST likely to support globalization?
   a. A CEO of a major corporation
   b. A leader of the Republican party
   c. A leader of the Democratic party
   d. A member of a labor union
   e. An economist

5. Middle-ground globalization scholar Dani Rodrik agrees that openness to the global economy can be highly beneficial, but to make it work, countries must invest in...
   a. Civil and political liberties
   b. Education
   c. Social insurance
   d. A and B above
   e. A, B, and C above

6. Where are most wars fought today?
   a. Between countries
   b. Globally
   c. In Latin America
   d. In the Middle East
   e. Within countries

7. With what other discipline can geography be readily compared?
   a. Sociology
   b. Political Science
   c. History
   d. Economics
   e. Anthropology

8. All of the following topics are covered under "human-environment interaction," EXCEPT...
   a. Does Brazil's "sustainable city," Curitiba, live up to its billing?
   b. How do the politics, economies, and environments of Europe compare to those of East Asia?

c. How have humans adapted to life in the arid lands of North Africa?

d. What are the local and global effects of deforestation in the Amazon rain forest?

e. What will happen to the plants and animals living in and around China's Three Gorges Dam as construction of the dam continues?

9. Which of the following states would be most likely to be included in the vernacular region of the "Midwest" in the United States?
a. Arizona   b. Georgia   c. Indiana   d. Maine   e. Oregon

10. The vast majority of current population growth occurs in four world regions. Which of the following is NOT one of these world regions?
a. South Asia   b. Latin America   c. East Asia   d. Europe   e. Africa

11. Which measure of population is produced by subtracting the number of deaths in a given year from the number of births in that same year?
a. Total Fertility Rate (TFR)
b. Percent of population under age 15
c. Rate of Natural Increase (RNI)
d. Percent of Population over age 65
e. Migration

12. What does a population pyramid with a wide base and a narrow top tell us about the growth rate of the population represented by the pyramid?
a. The population is experiencing rapid growth
b. The population is experiencing slow growth
c. The population is experiencing no growth
d. The population is experiencing negative growth
e. Population pyramids cannot tell us anything about the growth rate

13. Which of the following countries is NOT one of the top destinations for international migrants?
a. United States   b. France   c. Germany   d. Canada   e. China

14. When the modern country of Turkey was established in the early 1900s, its leader (Ataturk) created the Turkish Linguistic Society and charged it with purging the language of foreign phrases. This is an example of which of the following practices?
a. Cultural assimilation
b. Cultural hybridization
c. Cultural imperialism
d. Cultural nationalism
e. Cultural syncretism

15. A high birthrate usually goes along with which of the following statistics?
    a. High female illiteracy
    b. High GNP per capita
    c. High life expectancy at birth
    d. Low under-age-5 mortality
    e. All of the above

Answers: 1-A; 2-B; 3-A; 4-D; 5-E; 6-E; 7-C; 8-B; 9-C; 10-D; 11-C; 12-A; 13-E; 14-D; 15-A

Chapter 2
THE CHANGING GLOBAL ENVIRONMENT

LEARNING OBJECTIVES
- The purpose of this chapter is to provide an overview of Earth's environmental systems—geology, climate, hydrology, and vegetation—to set the scene for better understanding the environmental geography of the 12 world regions covered in the following chapters.
- This chapter introduces a variety of concepts and global relationships that the student must fully understand in order to synthesize the information in individual regional chapters to follow.
- It also introduces the elements of the physical geography of each region that students will see in each succeeding chapter: land structures, climate, water, and vegetation.
- This chapter introduces and explains a number of foundation concepts, including the dynamic geology of Earth, including earthquakes and volcanoes; the forces that propel and influence global climates; the dilemmas associated with Earth's most plentiful—and yet most critical—resource, water; and their interrelationships with plants and animals. This chapter also places humans in this symbiotic system.
- When the student has completed this chapter, he or she should have a solid foundation of knowledge on Earth's physical characteristics that will help him/her to assimilate the knowledge about regions that will be presented later in the book. Among the specific concepts and models that the student should understand are the following:

    · Plate tectonics
    · Geologic hazards, earthquakes and volcanoes
    · Climate regions
    · Climate change and global warming
    · Water cycles
    · Human impacts on the environment
    · Deforestation and desertification
    · The Green Revolution

CHAPTER OUTLINE
I. Geology and Human Settlement: A Restless Earth
- A. Geology shapes the fundamental form of Earth's surface
    1. Human-environment interaction
    2. Geology shapes landscapes
- B. *Plate tectonics:* A geophysical theory that states that the surface of Earth is made up of a large number of geological plates that move slowly across the surface of the planet (Fig. 2.2)
    1. Three major zones: core, mantle, outer crust
    2. Cooling of core results in heat exchange that causes exchange of materials through mantle
    3. *Convection cells:* large areas of very slow-moving molten rock within Earth

4. *Tectonic plate:* similar to a fractured jigsaw puzzle (Fig. 2.3)
5. *Convergent plate boundary:* a point on Earth where two plates are being forced together by convection cells deep within Earth (e.g., San Andreas fault running north-south in coastal California)
6. *Subduction zone:* a region where one tectonic plate dives below another; characterized by deep trenches where the ocean floor has been pulled downward by tectonic movements
7. *Divergent plate boundary:* a region where tectonic plates move away from each other in opposite directions
   Rift valleys are divergent plate boundaries that form deep depressions (e.g., the Red Sea)
8. *Pangaea:* Earth's original single supercontinent, centered on present-day Africa
C. Geologic Hazards: Earthquakes and Volcanoes
   1. Generally, each year over 100 earthquakes kill thousands and cause damage; earthquakes are unpredictable
   2. Earthquakes and volcanic eruptions are found along most tectonic plate boundaries

II. Global Climates: An Uncertain Forecast
A. Climate links the people of the Earth together in a global economy, providing opportunity to some, hardship to others, and challenges to all as we struggle to feed the world
B. Climatic Controls: five main factors influence meteorological conditions
   1. Solar energy: incoming solar energy *(insolation)* is absorbed by land and water, and heats Earth's lower atmosphere, much as a greenhouse, and is called the "greenhouse effect." Without some of it, Earth would be too cold for human habitation
   2. Latitude: because of Earth's curvature, highest levels of insolation occur at the equator (0 degrees latitude)
   3. Interaction between land and water: land heats and cools faster than water does; places in the middle of a continent have the greatest temperature extremes (heat and cold): *continentality. Maritime climates,* close to oceans, have more moderate temperatures, and are often moist
   4. Global pressure systems: Uneven heating of Earth and arrangement of oceans and continents creates high- and low-pressure cells, leading to different weather patterns (including storms) on Earth.
   5. Global wind patterns: Pressure systems produce wind because air (wind) flows from high-pressure areas to low-pressure areas. These winds help carry weather patterns from one place to another
C. World Climate Regions
   1. *Weather:* short-term, day-to-day (or even hourly) expression of atmospheric processes (rainy, hot, etc.)
   2. *Climate:* A long-term view of the weather of a region, based on the compilation and statistical averaging of data on temperature, pressure, precipitation, humidity, etc. for a period of at least 30 years

3. *Climate region:* An area within which similar climatic conditions prevail (Fig. 2.12)
4. *Climograph:* Graphs of average high and low temperatures and precipitation for the 12 months of the year (Fig. 2.12)

D. Global Warming
1. Human activities connected with economic development and industrialization seem to be changing the world's climate, especially raising its average temperature
2. *Anthropogenic:* human-caused phenomena; refers to the human-caused emissions associated with global warming
3. Causes of global warming include: (1) carbon dioxide from burning fossil fuels (>50 percent greenhouse gases); (2) chlorofluorocarbons from aerosol sprays and refrigerants (25 percent); (3) methane from burning vegetation (rain forests), by-product of cattle and sheep digestion, and (4) leakage of pipelines and refineries (15 percent); nitrous oxide from chemical fertilizers (5 percent)
4. Effects of global warming include increase in average global temperature by 2–4 °F by 2030; climate change may cause shift in agriculture areas, decrease in grain production, increase in sea level (thermal expansion of oceans, ice cap melting)

E. Globalization and Climate Change: The International Debate on Limiting Greenhouse Gases
1. Rio de Janeiro, Brazil, 1992, first international agreement on limiting greenhouse gases signed

2. Worst offenders: United States, Japan, India, China
3. Kyoto, Japan, 1997, meeting raised key issues: United States concerned that reducing emissions will slow economy; European Union is in compliance; developing regions do not want to sign an agreement that will threaten their economic prospects; economy or environment. In 2005, after over 100 countries accounting for 55 percent of greenhouse gas production ratified Kyoto Protocol, it went into effect. Russia's ratification was key; U.S. did not ratify

III. **Water on Earth: A Scarce and Polluted Resource**
A. Fresh Water Is Unevenly Distributed: about 40 percent of the Earth's people live in arid or semiarid lands
B. Global Water Budget
1. 70 percent of Earth's surface is covered by water
2. Only 3 percent of Earth's water is fresh
3. 99 percent of the fresh water is locked up in polar ice caps and mountain glaciers (only 1 percent of the 3 percent of fresh water is available!)
4. Pollution further cuts down on the amount of usable freshwater available
5. *Water stress:* uses data based on the amount of fresh-water available on a per capita basis in different parts of the world to describe and predict where water resource problems will be greatest (Africa, Northern China,

India, Southwest Asia, Mexico, part of Russia) (Fig. 2.167)

C. Flooding
1. Floods cause more deaths than other natural disasters (50 percent of deaths from natural disasters)
2. Rising populations and settlements in floodplains and other similar regions increases deaths from flooding
3. Deforestation increases the severity of flooding

IV. **Human Impacts on Plants and Animals: The Globalization of Nature**
A. Biomes and Bioregions
Biome: term used to describe a grouping of flora and fauna into a large ecological region
Bioregion: a spatial unit or region of local plants and animals adapted to a specific environment, such as a tropical savanna (in *Globalization and Diversity,* bioregion is used as a synonym for "biome")
1. Includes naturally occurring flora and fauna
2. Shaped by our domestication — and modification—of these plants and animals
3. Today bioregions may be home to multinational conglomerates as well as to local plants, animals, and peoples
4. Biome and Bioregion are used interchangeably in "Diversity Amid Globalization."

B. Tropical Forests and Savannas
1. Mostly in equatorial regions
2. High average temperatures
3. Long days year round
4. High amounts of rainfall
5. Usually three tiers of vegetation (200 ft., 100 ft., forest floor)

6. Most nutrients stored in living plants, rather than in soil; soil not fertile
7. Not suited to agriculture

C. Deforestation in the Tropics
1. Deforestation in Southeast Asia is occurring at a rate 3 times faster than in the Amazon; if the pace continues, Southeast Asia's forests could all be gone in 15 years.
2. Causes of deforestation: globalization of timber industry; replacement of forests with cattle farms to satisfy world demand for beef; settlement areas for rapidly growing population in developing regions—often all three in a single place!

D. Deserts and Grasslands
1. One-third of Earth's land is true desert, receiving <10 inches of rainfall per year
2. Grasslands may be *prairies* (longer denser grasslands, such as those in the midsection of Canada and the United States); or *steppes* (shorter, less dense grasslands, such as those in Central Asia, Russia, and Southwest Asia)
3. *Desertification:* spread of desert-like conditions into grassland areas; may be caused by overgrazing, poor cropping practices, and buildup of salts in soils from irrigation

E. Temperate Forests: large tracts of forests found in middle and high latitudes
1. Two types of trees dominate: softwood coniferous or evergreen and deciduous trees that drop their leaves in winter
2. Many of these forests have been cut down to make farmland

3. Logging pressures exist, too

V. Food Resources: Environment, Diversity, Globalization

A. If the human population continues to grow at expected rates, food production must be doubled by 2025 to provide each person with a basic subsistence diet

B. Industrial and Traditional Agriculture
   1. *Industrialized agriculture* is practiced in about 25 percent of the world's farmlands; uses large amounts of energy and chemicals from fossil resources; and often relies on irrigation
   2. *Traditional agriculture* is practiced by half the people on earth; *intensive agriculture:* high intensity production using many inputs (i.e., industrial farming or high intensity rice farming); *extensive agriculture:* low intensity production using large amounts of land (i.e., ranching)
   2. *Subsistence agriculture:* farming that produces only enough crops or livestock for a farm family's survival

C. The Green Revolution: interconnected processes to increase world food production, implemented since 1950
   1. Stage 1 (1950–1970) has three steps: (1) change from traditional mixed crops to monocrops, or single fields, of genetically altered, high-yield rice, wheat, and corn seeds; (2) intensive applications of water, fertilizers, and pesticides; (3) increasing intensity of agriculture by reducing the fallow or resting time between seasonal crops
   2. Stage 2 (since 1970s) has emphasized new strains of fast-growing wheat and rice specifically bred for tropical and subtropical climates that permit two or even three crops per year, rather than just one
   3. Environmental costs of Green Revolution: (1) 400 percent increase in agricultural use of fossil fuels in past few decades (10 percent of world's oil); (2) damage to wildlife and habitat from diversion of streams to farms; (3) pollution (especially of water) caused by pesticide and chemical fertilizer runoff from farms; (4) air pollution from factories making agricultural products
   4. Social costs of Green Revolution: (1) traditional farmers cannot compete against Green Revolution farmers; (2) high cost of Green Revolution strategies keeps many farmers out; (3) economic stratification of farmers creates economic, social, political tension

D. Problems and Projections
   1. Local issues, such as poverty, civil unrest, and war keep people from getting adequate food and nutrition
   2. Political problems are usually more responsible for food shortages than natural disasters (including drought); food distribution is politicized
   3. Globalization is causing dietary preferences to change worldwide: (1) more meat in diets (currently two-thirds are primarily vegetarian); (2) changing cultural tastes and values

4. Africa, South Asia are areas of greatest concern regarding food

VI. Conclusion
   A. Some Environmental Change Occurs because of Natural Forces
   B. Other change is caused by humans
   C. Global Environmental Science Is a New Area of Study

   D. Globalization Both Helps and Hinders World Environmental Problems
      1. A growing number of countries are willing to sign international agreements to improve environment (help)
      2. Superheated economic activity caused by globalized economy contributes greatly to world environmental problems

PRACTICE MULTIPLE CHOICE QUIZ

1. Why is the theory of plate tectonics important?
   a. It explains and describes the inner workings of Earth
   b. It explains and describes many of Earth's surface landscape features
   c. It gives clues about the world distribution of earthquakes and volcanoes
   d. A and B above
   e. A, B, and C above

2. Where does the heat exchange described in plate tectonic theory take place?
   a. At the convergent plate boundary
   b. At the divergent plate boundary
   c. In convection cells
   d. In the subduction zone
   e. In the center of the tectonic plates

3. Which of the following natural hazards is hardest to predict?
   a. Earthquakes   b. Hurricanes   c. Flooding   d. Blizzards   e. Volcanic eruptions

4. Which of the following statements about volcanoes is FALSE?
   a. The loss of life from volcanoes is a fraction of that from earthquakes
   b. Volcanoes are found along most tectonic plate boundaries
   c. Volcanoes can cause major destruction
   d. Volcanic eruptions cannot be predicted
   e. Volcanoes provide some benefits

5. What is the major factor that works to control Earth's climate?
   a. Latitude
   b. Insolation
   c. Interaction between land and water
   d. Global pressure systems and wind patterns
   e. All of the above

6.   Which of the following cities is most subject to continentality in its climate?
       a. Duluth, Minnesota
       b. Miami, Florida
       c. San Diego, California
       d. Seattle, Washington
       e. Virginia Beach, Virginia

7.   What does the Köppen system describe?
       a. Climate   b. Elevation   c. Landforms   d. Population   e. Soils

8.   Which of the following statements about the Greenhouse Effect is FALSE?
       a. Emissions in the lower atmosphere are increasing the greenhouse effect.
       b. Carbon dioxide accounts for more than half of the human-generated greenhouse gasses
       c. Natural greenhouse gasses have varied over long periods of geologic time
       d. The composition of greenhouse gasses has changed dramatically because of industrialization
       e. The earth would be far better of if we could eliminate the greenhouse effect

9.   As of March 2008, what is the only industrialized country that has NOT ratified the Kyoto Protocol?
       a. Australia   b. Brazil   c. Norway   d. Russia   e. United States

10.   What is the goal of the Kyoto Protocol?
       a. Develop a world system to predict tsunamis
       b. Implement a world system to increase crop yields
       c. Improve world water quality
       d. Limit greenhouse has emissions
       e. All of the above

11.   How have humans changed the natural patterns of plants and animals on earth?
       a. Plowing grasslands
       b. Burning woodlands
       c. Cutting forests
       d. Hunting animals
       e. All of the above

12.   Cattle from which of the following countries are now being raised on grassland pastures created by cutting tropical forests?
       a. Germany
       b. India
       c. Japan
       d. New Zealand
       e. Russia

13. Fully one-third of the world's land qualifies as this type of landform and climate region.
    a. Tropical rainforest
    b. Desert
    c. Mountain
    d. Prairie
    e. Steppe

14. Traditional intensive farming generally requires large inputs of this resource.
    a. Fertilizer
    b. Herbicides
    c. Labor
    d. Pesticides
    e. Irrigation

15. Which of the following is NOT an environmental cost of the Green Revolution?
    a. Stratification of society based on wealth and poverty
    b. Pollution of rivers and water sources
    c. Increased air pollution from factories producing agricultural chemicals
    d. Damage to wildlife
    e. Destruction of habitat

Answers: 1-E; 2-C; 3-A; 4-D; 5-E; 6-A; 7-A; 8-E; 9-E; 10-D; 11-E; 12-B; 13-B; 14-C; 15-A

Chapter 3
NORTH AMERICA

LEARNING OBJECTIVES
- This chapter is extremely important because it provides students with their first opportunity to use the introductory materials of this book to help them assimilate knowledge about a region of the world. The region covered by this chapter, North America, is one with which students should have some familiarity.
- In particular, this chapter provides the students with the first opportunity to apply the concepts of globalization and diversity to help improve their understanding of a region with which they are already familiar. This pattern sets the tone for the remainder of the book.
- This chapter introduces North America, a region of great diversity, and provides detailed discussion of the evolution of cities as well as the structure of metropolitan areas.
- This chapter provides a first opportunity to integrate introductory materials of this book with a region of the world. These introductory materials include human and physical geography concepts from Chapters 1 and 2.
- At the conclusion of this chapter, the student should be familiar with the physical, demographic, political, and economic characteristics of North America.
- The student should also be able to place his or her place of residence in North America within the context of the materials presented in the chapter.
- In addition, the student should understand the following concepts and models:

    · Megalopolis
    · Historical evolution of a U.S. city (stages of intraurban growth)
    · Urban structure models (concentric zones and urban realms)
    · Federalism
    · Sectoral transformation
    · North American Free Trade Agreement (NAFTA)

CHAPTER OUTLINE
I. Introduction
North America is comprised of two countries, the United States and Canada. This region is rich in resources, including both high-quality farm land and minerals. North America has a diverse cultural base, including indigenous peoples and immigrants from all over the world. Both the United States and Canada have evolved from British colonies and established democratic governments. The region is in the final stage of the demographic transition, and its economy is based largely on service and information sectors. Transportation and communication are very well developed. This is a region of great wealth, which nonetheless exhibits some disparities in the distribution of this wealth. The region is an example of a *postindustrial economy*. Globalization is evident throughout North America, with evidence of many other

cultures; and cultures from this region have had a strong impact elsewhere around the world.

II. Environmental Geography: A Threatened Land of Plenty

   A. The Costs of Human Modification (agriculture, population)

      1. Transforming Soils and Vegetation: Europeans brought new species (wheat, cattle, horses); settlers cut millions of acres of forest, and replaced grasslands with grain and forage crops; urbanization further changed the land, with 2 million+ acres of farmland converted to urban uses each year; soil erosion is another problem

      2. Managing Water: North Americans use lots of water; city dwellers use 175 gal/person/day; indirectly, through food and industrial consumption, we use about 1,400 gal/person/day; water quality and quantity are problems; Ogallala Aquifer has dropped 100 feet in 50 years; U.S. has Clean Water Act; Canada has Green Plan

      3. Altering the Atmosphere
Urban heat islands occur around cities, which are usually hotter than their surrounding hinterlands because human activities raise ambient air temperatures (thus the term *urban heat island*)

      4. Air pollution increases precipitation in cities

      5. Acid rain occurs when airborne pollutants (especially sulfur and nitrogen) mix with moisture in the air and come down as precipitation; usually originates in industrial areas; once in the air, winds carry pollutants a long way; a chemical reaction occurs, making a weak acid by combining molecules from the pollutants and the moisture in the air, creating an acidic rain; the acid rain often falls in areas that did not produce the original pollutants; acid rain kills vegetation, wildlife, and can dissolve buildings

   B. The Price of Affluence: toxic waste (and waste dumps; poor air quality; wild lands lost to development; United States and Canada work to improve environment

   C. A Diverse Physical Setting

      1. Gulf-Atlantic Coastal Plain (southern New York to Texas) has good farmland and accessibility; was the site of early European settlement and urbanization

      2. Piedmont, Appalachian Mountains, and interior highlands (including Appalachians, Ozarks, and Ouachita Plateau)—rugged terrain, less densely settled

      3. Interior lowlands (west central Canada to coastal lowlands near Gulf of Mexico, including Great Lakes and lower Ohio River Valley) is the largest landform region, has glacial and depositional plains, good soils, gentle slopes for farming

      4. Rocky Mountains (running from Alaska to northern New Mexico), includes high mountains, spectacular fault-blocks, mineral riches, and stunning views

      5. Pacific Mountains and valleys (Southeast Alaska, British Columbia, Washington, Oregon, California) active volcanoes, earthquakes, good harbors, and

good farming where water is available

D. Patterns of Climate and Vegetation (Fig. 3.8)
1. Great variation in climate and vegetation because of this region's size, latitudinal range, varied terrain, and oceans
2. South of Great Lakes: long growing season, deciduous forests, commercial agriculture
3. Great Lakes and North: boreal forest (conifers)
4. Hudson Bay and North: tundra
5. Prairie regions to the west have drier climates and grasslands
6. West Coast: Marine West Coast climate in San Francisco; Mediterranean in southern California

E. North American and Global Climate Change
1. High latitude and alpine climates are thawing, increasing erosion, affecting migrating polar bears and whales; southern species are moving north; longer growing seasons, but more tornadoes
2. Influences on physical geography include glaciers in Canada and eastern United States; ocean currents and temperatures that drive El Nino and La Nina; drought cycles on the Great Plains
3. Cycles may change in the future
4. People are moving to some of the most hazardous regions (with landslides, fire hazards, etc.)

III. Population and Settlement: Re-fashioning a Continental Landscape

A. Modern Spatial and Demographic Patterns (Fig. 3.10)
1. North America has more than 335 million people (302 million in the U.S., 33 million in Canada)
2. Megalopolis (mega = great, big, huge; polis = city): the largest settlement agglomeration in the United States, extending from Boston, Massachusetts, to Washington, D.C., (including New York City, Baltimore, and Philadelphia)

B. Occupying the Land
1. Indigenous people (Native Americans in the United States, First Nations in Canada) lived throughout North America for the past 12,000 years (3.2 million in continental U.S., 1.2 million in Canada, Hawaii, and Greenland)
2. European diseases, settlement patterns, and warfare decimated these peoples
3. Three stages of European Settlement of North America
   Stage 1: European colonial footholds, mostly on the East Coast from 1600–1750, including French, English, Dutch, Spanish, and enslaved Africans
   Stage 2: infilling better farm land in eastern half of North America, 1750–1850, populated Upper Ohio and Tennessee Valleys, interior lowlands, then Midwest and interior South; Canadian settlement slower
   Stage 3: acceleration of settlement, most of the region's farmlands, from 1850–1910; these settlers included both immigrants from abroad and North American-born migrants traveling to the West and West Coast for mineral (gold) rushes and other opportunities

22

C. North American Migration
1. Westward-moving populations: by 1990, more than half the U.S. population lived west of the Mississippi River.
   Trails, railroad tracks, highways facilitated westward movement. Today's fastest-growing: Texas, Arizona, Nevada, Alberta, and British Columbia, where high-tech jobs are a pull force; net outmigration from California
2. Black exodus from the South. Original African concentration—southern plantations. After emancipation, most African Americans stayed in the South. Migrants sought industrial jobs in North (1910–20 and 1940–60). Today many African Americans are returning to the South (reverse migration)
3. Rural-to-urban migration
   From farms to industry and business in the city. Today North America is 75 percent urban. Mechanization cut many farm jobs; business, industry drew people to cities. *Counter-urbanization:* people leave large cities and move to smaller towns and rural areas
4. Growth of the Sunbelt south. Fastest growing region since 1970. Jobs, strong economy, modest living costs are attractive. *Counterurbanization* trend: people find or create jobs in affordable smaller cities and urban areas
   Lifestyle migrants: move to places that offer amenities and are perceived as avoiding city problems
D. Settlement Geographies: The Decentralized Metropolis

1. *Urban decentralization:* when metropolitan areas sprawl in all directions and suburbs have characteristics of traditional downtowns
2. Historical evolution of the city in the United States (growth of the American city or stages of Intraurban growth) (Fig. 3.16): Transportation has had a strong influence on the size and shape of cities; four stages: (1) walking-horse-car, pre-1888; (2) electric streetcar, 1888–1920; (3) recreational automobile, 1920–1945; (4) freeway, 1945–present
3. *Concentric zone model*
   1=Central business district
   2=Zone of transition
   3=Zone of workers' homes
   4=Zone of better residences
   5=Commuters' zone
4. *Urban realms model:* new suburban growth with a mix of retail, industrial parks, office complexes, entertainment; also called *edge cities*
5. Consequences of Sprawl: Resources flee city for suburbs; farmland converted to suburbs; longer commutes, city problems move to suburbs
   Gentrification: movement of upscale individuals to run-down inner city areas; this often displaces lower-income residents, but rehabilitates deteriorated inner cities
   Suburban downtowns: similar to edge cities; suburbs are becoming full-service urban centers unto themselves, including retail, business, education, jobs, etc., as well as housing.

E. Settlement Geographies: Rural North America
    1. *Township-and-range survey system:* introduced in United States 1785 to lay out land in unincorporated areas; caused squared-off pattern oriented along the cardinal directions (north, south, east, west); Canada has a similar system
    2. Railroads were crucial to opening interior North America to settlement from coast-to-coast
    3. Family farms have been replaced by corporate farms
    4. Population drain in places like rural Iowa, eastern Montana, southern Saskatchewan

IV. Cultural Coherence and Diversity: Shifting Patterns of Pluralism
    A. North America's cultural geography is globally dominant and internally pluralistic
    B. The Roots of a Cultural Identity
        1. Early dominance of British culture (U.S. and Canada); then consumer culture after 1920 provides common experience
        2. *Ethnicity:* exists when a group of people with a common background and history identify with one another, often as a minority group within a larger society; both Canada and the United States have significant ethnic communities
    C. Peopling North America: five phases of migration to United States (first three phases dominated by Europeans)
        1. Phases of migration to United States (Fig. 3.19): *Phase 1* (before 1820)—English and Africans; *Phase 2* (1820–1870)—Irish and Germans;

Phase 3 (1870–1920)—Eastern Europeans (Polish, Russian, Austro-Hungarians); *Phase 4* (1920–1970)—Canadians, Latin Americans dominate, with sharp drop in number; *Phase 5* (1970–present)—Asians, Latin Americans increase in numbers
        2. The Canadian pattern is similar to that of the United States; large French influence, especially in Quebec
        3. *Cultural assimilation:* the process by which immigrants are absorbed into the larger host society
    D. Culture and Place in North America: culture and ethnic identity are often strongly tied to place (Fig. 3.23)
        1. Persisting *cultural homelands* (a culturally distinctive nucleus of settlement in a well-defined geographical area; its ethnicity survives over time and stamps the landscape with an enduring personality)
        2. There are several cultural homelands in North America: French in Quebec (more than 80 percent of Quebecois speak French); Hispanic borderlands in southwest United States; Black Belt in southeast United States; Navaho Reservation in Four Corners area of United States.
        3. A Mosaic of Ethnic Neighborhoods (Fig. 3.25): ethnic geography of cities reflects economics there, and cultural patterns of migration and settlement; examples include Chinatowns; ethnic neighborhoods are common in large cities
        4. Patterns of North American religion: 60% Protestant; 24%

Roman Catholic and concentrations of many other religions (Orthodox Christians, Jews, Muslims, Buddhists, and Hindus) in urban areas especially; Latter Day Saints (Mormons) dominate in Utah and Idaho

E. The Globalization of American Culture
   1. North Americans: living globally—buying and using products from around the world, bringing culture to North America via entertainment, news media, and traveling abroad
   2. Migrants to the U.S. are learning English more rapidly
   3. The global diffusion of U.S. culture through action of the government (Marshall Plan, Peace Corps) and exports by private corporations (McDonalds, Coca-Cola, Disney, CNN)
   4. Challenges to U.S. cultural dominance: for example, Canada requires minimum level of Canadian content in media; France has "language police"

V. Geopolitical Framework: Patterns of Dominance and Division (Fig. 3.27)

A. Creating Political Space: United States and Canada have different political roots
   1. Uniting the states—United States broke cleanly and violently from Britain, then annexed western lands guided by *Manifest Destiny,* a policy that held that the United States should cover the land from east coast to west coast
   2. Assembling the provinces (Canada)—Canadian Confederation separated slowly and smoothly from Britain, grew in piecemeal fashion, while Anglo and French Canada parted in 1791, reunited in 1840

B. Continental Neighbors
   1. United States and Canada share 5525-mile-long border (longest unprotected international border in the world)
   2. United States and Canada have long history of cooperation (see Fig. 3.26, Geopolitical Issues in North America)
   3. United States and Canada are partners in the International Joint Commission (created in 1909 by Boundary Waters Treaty), which resolves cross-boundary issues involving water resources, transportation, environmental quality
   4. Canada, the United States, and Mexico are partners in the North American Free Trade Agreement (NAFTA), designed to reduce barriers to trade and capital investment among the three countries

C. The Legacy of Federalism
   1. Two types of state (country) government (1) *unitary:* power centralized at the national level; and (2) *federalism:* allocates considerable political power and autonomy to units of government under the national level
   2. Early federalism in United States and Canada: United States began with limited central authority and allocated all unspecified authority to states or individuals, while Canada reserved most powers for the central government
   3. Evolution of federalism: United States has evolved a more

powerful central government, while Canada has shifted toward more provincial (state) autonomy

D. Quebec's Challenge
1. French is the official language in Quebec
2. Autonomy, not secession, is the goal of French Canadians
3. Meech Lake Accord (1990) included strong guarantees of Quebec's status as a "distinct society"

E. Native Peoples and National Politics
1. Indian Self Determination and Education Assistance Act of 1975 (United States) began trend of greater Native American autonomy
2. Indian Gaming Regulatory Act of 1988 (United States) raised $25 billion in 2006 from gambling on reservations
3. In Canada, Native Claims Office (1975) allocated millions of acres of land to native groups
4. Canada created Canada's newest territory, Nunavut, from the eastern part of Northwest Territories in 1999; 85 percent of Nunavut's 30,000 people are Inuit

F. Politics of U.S. Immigration
1. Questions about how many immigrants should be allowed
2. What is a fair policy for undocumented workers?

G. A Global Reach
1. U.S. role as extends outside its border (Monroe Doctrine-1824)
2. United States took isolationist stance in 1920s and 30s
3. Since World War II, United States has played important role around the world, including NATO and OAS, and wars in

Korea, Vietnam, involvement in Panama, Nicaragua, Kosovo
4. Troops currently in Afghanistan, Iraq, elsewhere (Fig. 3.30)

VI. **Economic and Social Development: Geographies of Abundance and Affluence**
A. An Abundant Resource Base (Fig. 3.31): many and varied resources result in high levels of economic development
B. Opportunities for Agriculture
1. High productivity in farming: only 2.1 percent of U.S. citizens, 2.7 percent of Canadians work as farmers, but produce ample food
2. Commercial farms taking over family farms: in the United States, there fewer farms than in 1975; average farm size rose
3. Canada is the world's largest exporter of timber, pulp, newsprint
C. Industrial Raw Materials
1. North America is a net importer of petroleum, producing 12 percent of world's oil, using more than 25 percent
2. United States imports petroleum, natural gas from Canada
3. United States has 27 percent of the world's coal (400-year supply at current levels of use)
4. Other extractive resources include copper, lead, zinc, gold, silver, nickel, iron ore
D. Creating a Continental Economy
1. Connectivity and economic growth: *connectivity*—how well cities are linked by transportation and communication; improvements in transportation and communication improved connectivity and fostered economic growth

2. *The sectoral transformation:* the evolution of a nation's labor force from one highly dependent on the primary sector, to one that has more employment in other sectors; in the United States and Canada, more than 70 percent of workers are in the tertiary and quaternary sectors
   Primary sector: natural resource extraction (mining and farming)
   Secondary sector: manufacturing or industrial
   Tertiary sector: services
   Quaternary sector: information processing

E. Regional Economic Patterns
   1. *Location factors:* the varied influences that explain why an economic activity is located where it is; examples of location factors are (1) proximity to natural resources, (2) connectivity, (3) productive labor, (4) market demand, (5) capital investments, (6) government spending, and (7) access to innovation and research
   2. Major manufacturing regions (Fig. 3.31): megalopolis, the Midwest, Sunbelt areas, West Coast
   3. *Agglomeration economies:* where many companies with similar and often integrated manufacturing operations locate near one another

F. North America and the Global Economy
   1. Patterns of trade: North America is a major importer and exporter; U.S. exports include software, entertainment products, cars, aircraft, telecommunications equipment, computers, financial and tourism services, food products; Canada's exports include grain, energy, wood, manufactured goods
   2. Patterns of global investment: foreign capital investments in stocks, bonds, and foreign direct investment (FDI); Japanese and German car factories in North America, investment in U.S. real estate. United States is the largest destination of foreign investment in the world.
   3. Doing business globally: many United States firms have established operations abroad (e.g., Coca-Cola, Intel, Procter & Gamble, General Electric), including selling products and operating factories overseas whose products are sold in North America
   4. Three shifts: (1) United States firms are becoming more globally integrated; (2) non-U.S. MNCs are buying United States and European firms and assets; (3) many MNCs are investing outside of North America (Africa, Southeast Asia, etc.)

G. Persisting Social Issues
   1. Wealth and poverty: high personal income in North America
   2. But 13 percent of United States residents and 18 percent of Canadians are poor, with regional and ethnic variations
   3. *Digital divide:* poor and underprivileged groups have much less access to the Internet than the wealthy

H. Twenty-First Century Challenges
   1. United States, Canada social indicators compare favorably

2. Concerns remain: declining availability of good jobs; need for more investment in education; gender gap in earnings and opportunities; health care and aging, chronic diseases; racial issues

distribution of wealth. North American culture has spread throughout the world, and because of several major waves of international immigration, itself has a diverse cultural heritage. Advances in technology put this region (along with Japan and Europe) at the forefront of world economic and political interactions

VII. Conclusion
North America is a wealthy region that continues to exhibit disparities in the

PRACTICE MULTIPLE CHOICE QUIZ

1. Which of the following statements about North America is FALSE?
   a. It has the highest rates of resource consumption on Earth
   b. Its population is about 335 million people
   c. It is home to 30% of the world's population and consumes 30% of its resources
   d. It is one of the world's most affluent regions
   e. It is rich in natural resources

2. How did European settlers shape the ecology of North America?
   a. They brought new animals, including cattle and horses
   b. They introduced new plants, including wheat
   c. They removed forest cover from millions of acres of land
   d. They replaced natural grasslands with non-native grain and forage crops
   e. All of the above

3. Which of the following statements about the Ogallala aquifer is FALSE?
   a. Regular rainfall keeps it from dropping more than a few feet
   b. It was formed during the last Ice Age
   c. It lies beneath the Great Plains
   d. It is the largest aquifer in North America
   e. It irrigates about 20% of U.S. cropland

4. Which of the following regions of North America is considered to be among its "most toxic localities?"
   a. Canada's Hudson Bay
   b. Michigan's Upper Peninsula
   c. Nunavut's Arctic Circle coastline
   d. Texas and Louisiana Gulf Coast
   e. Wyoming's mountains

5.   Only two countries in the world consume more water than the United States. What are they?
       a. Australia and New Zealand
       b. Germany and Great Britain
       c. India and China
       d. Russia and Japan
       e. Nigeria and Brazil

6.   Where are many producers of acid rain in North America clustered?
       a. New England
       b. The Desert Southwest
       c. The Midwest and Southern Ontario
       d. The Rocky Mountain States and Provinces
       e. The Pacific Northwest

7.   Which of the following cities lies within the original Megalopolis?
       a. Baltimore, Maryland
       b. Kansas City, Missouri
       c. Salt Lake City, Utah
       d. Phoenix, Arizona
       e. Montreal, Quebec

8.   Where were the earliest colonial settlements in North America located?
       a. Along the current U.S. border with Mexico
       b. Around the Great Lakes
       c. In the central states
       d. In the Pacific Northwest region of the U.S. and Canada
       e. Within coastal regions of eastern North America

9.   The most persistent regional migration trend in North America has been movement…
       a. because of pull factors
       b. from the south to the north
       c. from urban areas to rural areas
       d. to sunbelt states
       e. west

10.   Which of the following statements about African Americans in North America is FALSE?
       a. After emancipation, few former slaves continued to work in the south
       b. At the end of the American colonial period, African Americans made up the majority of
            the population in southern states of the U.S.
       c. Most African American migrants in the 1900s settled in northern cities
       d. Originally, African Americans were concentrated in the plantation South
       e. Today, many African Americans are migrating to the South

11.   Today, settlement landscapes of North American cities display the consequences of…
       a. Concentric zones
       b. Counterurbanization

c. Urban decentralization
d. A and B above
e. A, B, and C above

12. Which of the following is a cultural homeland in Canada?
   a. Acadiana
   b. Black Belt
   c. Hispanic Borderlands
   d. Navajo Reservation
   e. Nunavut

13. Which of the following statements about the relationship between the U.S. and Canada is FALSE?
   a. Agricultural competition sometimes creates tension between the two neighbors
   b. Canada and the U.S. joined to clean up the Great Lakes
   c. The U.S. is a member of NATO but Canada is not
   d. The two countries share a border that is more than 5000 miles long
   e. Trade relations between these neighbors is very important

14. Which part of North America is now its leading region of manufacturing exports?
   a. Gulf Coast industrial region
   b. Megalopolis
   c. Piedmont manufacturing belt
   d. Silicon Valley
   e. Southern Ontario

15. Which of the following locations in North American is marked by rural poverty?
   a. A ghetto on the west side of Chicago
   b. A retirement community in Florida
   c. A suburb of New York City
   d. Appalachia
   e. British Columbia

Answers: 1-C; 2-E; 3-A; 4-D; 5-C; 6-C; 7-A; 8-E; 9-E; 10-A; 11-C; 12-E; 13-C; 14-D; 15-D

Chapter 4
LATIN AMERICA

LEARNING OBJECTIVES
- This chapter is important because it provides students with their first opportunity to integrate introductory materials of this book with a relatively unfamiliar region of the world.
- In particular, this chapter provides the student with the first opportunity to compare one region with another, specifically, enabling him or her to compare Latin America with North America (Chapter 3).
- This chapter introduces Latin America, a region of great cultural diversity that bears the imprint of its history of colonialization.
- Within this region are found two of the world's most important physical features: the Amazon rain forest and the Andes Mountains.
- After exploring the materials in this chapter, the student should be able to locate major physical features in this region, identify the countries in the region, and understand how the physical geography of the region has contributed to the social and economic patterns we observe in the region.
- In addition, the student should understand the following concepts and models:
    - *El Niño, La Niña*
    - Grassification
    - Altitudinal zonation
    - Columbian exchange
    - Syncretic religions
    - Dependency theory
    - Growth poles
    - Maquiladora
    - Dollarization

CHAPTER OUTLINE
I. Introduction
Latin America is comprised of 17 countries with 527 million people, from Mexico on the north to the southern tip of South America. (Please note, this region does NOT include the islands of the Caribbean, which are the subject of the next chapter.) This region is rich in resources, including both high-quality farm land and minerals. It is also home to the world's largest tropical rain forest along the Amazon, as well as to the world's longest mountain chain, the Andes. Latin America bears the imprint of its colonizers from the Iberian Peninsula of Europe, the Spanish and Portuguese. Reminders persist of the trade of enslaved Africans, and the domination of the indigenous peoples. Latin America is also home to several megacities and primate cities, although urbanization is a mixed bag of modernity and poverty. Economic and political instability are slowly giving way to development. The Central American Free Trade Agreement (CAFTA) was signed in 2005.

II. Environmental Geography: Neotropical Diversity and Urban Degradation (Figs. 4.3, 4.12)

    A. Much of the region (but not all) lies in the tropics

    B. *Neotropics:* tropical ecosystems of the Americas that evolved in relative isolation and support diverse and unique flora and fauna

    C. Environmental Issues Facing Latin America
        1. Large size and low population density have moderated environmental damage
        2. Environmental movements in the region may help Latin America avoid degradation as the region develops

    D. Rural Environmental Issues: Declining Forests
        1. *Deforestation* (Fig. 4.3): affects countries including parts of Brazil, Mexico, Chile, and Central America; major causes include agriculture and ranching, settlement, commercial lumber
        2. Loss of biological diversity a consequence of deforestation (tropical rain forests cover 6 percent of Earth's landmass, account for 50 percent of species)
        3. *Grassification* (the conversion of tropical forest into pasture) has also contributed to deforestation

    E. Degraded Farmlands
        1. Introduction of hybrid crops is eroding genetic diversity of food crops
        2. Pollution from farm chemicals
        3. Soil erosion and fertility decline
        4. Industrial agriculture pushes subsistence farms to marginal areas
        5. Urban sprawl eating farmland

    F. Problems on Agricultural Lands
        1. Loss of native food crops
        2. Soil erosion, fertility loss

    G. Urban Environmental Challenges
        1. Modern urban challenges: include pollution, inadequate water, garbage removal; squatter settlements; industrial pollution
        2. Curitiba (Brazil) is a "green city," showing how planning can minimize the negative aspects of urbanization

    H. The Valley of Mexico
        1. The Valley of Mexico (around Mexico City), with its mild climate, fertile soils, and ample water, has now become so densely populated (18 million) that smog, water (quality and quantity), and subsidence are problems; poverty, government inaction make the problems worse
        2. Air pollution: smog is severe in Mexico City; *temperature inversion* occurs when a layer of warm air traps pollutants; happens most in winter, causing children to stay indoors
        3. Water resources: both scarcity and quality are a problem; about 70 percent of water used in the area comes from aquifers
        4. The sinking land: as Mexico City pumps water from its aquifer, the city is sinking (30 feet during the 20th century)

    I. Western Mountains and Eastern Shields
        1. Most important settlement regions are in shields, plateaus, and fertile intermontane basins
        2. The Andes: run 5,000 miles from northwest Venezuela to Tierra del Fuego (southern tip of South America); Andes are

relatively young, created by collision of oceanic and continental plates; active volcanoes and earthquakes common; 30 peaks higher than 20,000 feet

3. *Altiplano:* elevated plain (in Peru and Bolivia); Lake Titicaca is located here

4. Uplands of Mexico and Central America: most densely settled in the region; elevations from 4,000–8,000 feet; rich volcanic soil is fertile, good for farming; water now overused and polluted; many of its 40 volcanoes are active

5. The shields: Brazilian Shield largest, best for settlement and natural resources, includes Sao Paolo; Patagonia is steppe, with few settlements; has sheep raising and petroleum extraction occur here

J. River Basins and Lowlands
1. Amazon Basin: world's largest river system, by volume (drains 6.1 million square kilometers); second longest river; rainfall exceeds 60 inches/year (more in some places); 20 percent of all freshwater going into oceans comes from Amazon; navigable for over 2,000 miles (to Iquitos, Peru); sparse settlement

2. Plata Basin: second largest watershed in Latin America; includes Parana, Paraguay, and Uruguay Rivers; much of this area is economically productive through mechanized agriculture (soybeans); more arid regions support livestock

3. Orinoco Basin (3rd largest): Third largest in South America; flows through Venezuela and Colombia, and the llanos, a tropical grassland that supports cattle ranches; petroleum pumped here

K. Climate Patterns (Fig. 4.14)
1. Relatively little variation in temperature in the region because of tropical

2. Much more variation in precipitation; many places have distinct wet and dry seasons; tropical humid east of Andes; desert on Pacific coast of Peru and Chile, Patagonia, northern Mexico, and Bahia of Brazil; mid-latitude (hot summers, cold winters) in Argentina, Uruguay

3. *Altitudinal zonation:* As one goes to the top of a mountain (altitude increases), temperature drops, causing changes in vegetation, resulting in zones of vegetation related directly to altitude (Fig 4.16)
Environmental lapse rate: for every 1,000 feet in elevation, there is a drop in temperature of 3.5 degrees Fahrenheit
The zones:
Tierra caliente: hot land: Sea level - 3,000 feet (sugar cane, tropical fruits, lowland tubers, maize, rice, poultry, pigs, cattle)
Tierra templada: temperate land: 3,000–6,000 feet (coffee, maize, warm-weather vegetables, cut flowers, short-horn cattle)
Tierra fria: cold land: 6,000–12,000 feet (wheat, barley, maize, tubers, sheep, guinea pigs, llama, alpaca, vicuna)
Tierra helada: freezing land: above 12,000 feet (highland

33

grains and tubers, sheep, guinea pigs, llama, alpaca, vicuna

4. Latin America and Climate Change: Has 6% of world's population, produces 6% of emissions; the region has lower average energy use, more renewable energy, public transit; highland areas are most vulnerable to climate change, and will affect glacial runoff that provides much water to the region

5. *El Niño:* Warm Pacific current that usually arrives along coastal Ecuador and Peru in December, around Christmastime; usually occurs at 10-year intervals; produces torrential rains, leading to flooding in some places and drought in others

6. Drought: occasionally occurs in northeast Brazil; agricultural loss usually the worst problem; decreases hydropower, leading to rolling blackouts

III. Population and Settlement: The Dominance of Cities

A. Latin America had dramatic population growth in the 1960s and 1970s; the rate slowed in the 1980s
   1. Life expectancy increasing
   2. Shift to urban living

B. The Latin American City
   1. About 75 percent of Latin Americans live in cities; over 85 percent in Argentina, Chile, Uruguay, Venezuela; Spanish and Portuguese colonialists gave preferential treatment to city dwellers, thus providing motivation to move to cities
   2. High levels of *urban primacy*: condition in which a country has a *primate city* that is 3 to 4 times larger than any other city in the country; examples: Lima, Caracas, Guatemala City, Santiago, Buenos Aires, Mexico City; new megalopolises in Mexico, Brazil, and Argentina

   3. Urban form: Latin American cities have distinctive urban shape (Fig. 4.18) that reflects colonial origins and contemporary growth; these cities have a central business district (CBD), beltway-style highway, highest-quality residences and businesses in the city center, and along a "spine" extending to the beltway; squatter settlements exist beyond the beltway; *squatter settlements* are makeshift housing on land not legally owned or rented by urban migrants, usually on the outskirts of a rapidly growing city

   4. *Rural-to-urban migration:* movements of peasants from rural areas to the cities; these migrants often maintain ties with families in the countryside

C. Patterns of Rural Settlement: highly mechanized capital-intensive farming is somewhat common, and rural areas are linked with cities
   1. Rural landholdings: a system of large estates *(encomienda),* owned by absentee landlords, and using hired and slave labor was common; peasants did not own land, but worked for the estates; *latifundia:* the entrenched pattern of large estates associated with encomienda; *minifundia:* pattern associated with peasants farming small plots for subsistence;

34

agrarian reform: redistribution of land, giving land to peasants

2. Agricultural frontiers: opening up of land for peasants, while tapping unused resources and strengthening political boundaries; Brazil's Amazon Highway and development (including Brasilia) are an example

D. Population Growth and Movements
   1. Rapid growth throughout most of the 1900s, followed by slower growth since 1980
   2. European Migration: many Latin American countries encouraged migration; the greatest number of Europeans came to Argentina, Chile, Uruguay, and Southern Brazil from 1870–1930
   3. Asian Migration: Asians (Chinese, Japanese) came from 1870–1930, but in smaller numbers than Europeans; Peru had a Japanese-Peruvian president in the 1990s; South Koreans are newest wave
   4. Latino Migration and Hemispheric Change: pull factors (economic opportunity) and push factors (political turmoil, including civil wars) have caused movement within Latin America, or from Latin America to North America; *transnationalism:* many migrants maintain close contact with their home countries

IV. Patterns of Cultural Coherence and Diversity: Repopulating a Continent

A. Demographic Collapse
   1. There were many complex civilizations in Latin America before Europeans arrived
   2. The demographic toll: Latin America's population before

Europeans arrived was 54 million, 12 million more than in western Europe; after Europeans arrived, Latin America's population dropped to about 5 million (a 90 percent population loss!); European illnesses, warfare, forced labor, and collapse of the food production system all caused the demographic collapse

3. *The Columbian exchange:* an immense "biological swap" that occurred after Columbus came to the New World (term coined by Alfred Crosby); Europeans brought wheat, olives, grapes, pigs, cows, horses, sugar cane, coffee from Europe or their colonies in Africa and Asia; Amerindians provided starch crops (maize/corn, potatoes, hot peppers, tomatoes, pineapples, cacao, avocados); some Latin American Indians became expert horse riders

4. Indian survival: largest indigenous populations are in Mexico, Guatemala, Ecuador, Bolivia; land was the key factor in survival, and governments are providing *comarcas* (areas of land set aside for indigenous Amerindians over which native peoples exert political and resource control, unlike United States reservations)

B. Patterns of Ethnicity and Culture
   1. Complex system of racial caste under the Spanish, which intermarriage eventually caused to collapse; this system was replaced by a four-category system (Blanco—European; Mestizo—mixed ancestry;

Indio—Indian ancestry; Negro—African ancestry)
2. Enduring Amerindian Languages: about two-thirds of Latin Americans speak Spanish; one-third speak Portuguese; indigenous languages are spoken in small parts of the Central Andes, Mexico, and Guatemala
3. Blended religions; most countries are more than 90% Roman Catholic; others practice *syncretic religions,* which blend different belief systems (Animist/Christian; African/Catholicism)
4. Soccer: Brazil's Pele was influential in world soccer; soccer is an important sport in the region

C. Global Reach of Latino Culture
1. *Telenovelas:* popular nightly soap operas, exported globally
2. National identities: Each country of Latin America celebrates its unique identity and history; Latin music (including mariachi and samba) are becoming more popular outside the region; Latin literature is becoming more global

V. Geopolitical Framework: Redrawing the Map
A. Organization of American States (OAS) (1948): envisioned a neutral Pan-American vision of hemispheric cooperation, rather than dominance by the United States
B. Iberian Conquest and Territorial Division
1. *Treaty of Tordesillas (1493–94):* Catholic Pope interceded to divide uncharted lands between

Spanish (west half) and Portuguese (east half)
2. Revolution and independence: elites born in the Americas led the revolutions between 1810 and 1826, with existing political divisions the basis for countries
3. Persistent border conflicts Colonial territories not clearly delimited, especially in the sparsely populated interior; several wars resulted (including 1981 Falklands conflict between Britain and Argentina); today most conflicts are nagging diplomatic issues
4. The trend toward democracy: most countries in the region are free market democracies

C. Regional Organizations
1. Supranational organizations: governing bodies that include several states, usually to achieve a common goal
2. Subnational organizations: groups that represent areas or people within states; often ethnic or ideological (political), and can cause internal division
3. Trade blocks: regional alliances to foster internal markets and reduce barriers to trade; examples include: *Mercosur* (1991): Southern Cone Common Market (Brazil, Uruguay, Argentina, Paraguay) trade block, does a great deal of trade with the European Union; NAFTA (North American Free Trade Agreement) includes Mexico with the United States and Canada; CAFTA (2004) (Dominican Republic, Guatemala, El Salvador, Honduras, Nicaragua, Costa Rica)

4. Insurgencies and drug trafficking; examples: guerrilla groups (Shining Path/Sendero Luminoso in Peru, FARC/Revolutionary Armed Forces of Colombia) have controlled large areas of their countries through violence and intimidation; *Drug cartels:* powerful and wealthy organized crime syndicates, many centered in Colombia; United States helps fund efforts to reduce production of illicit drugs in Latin America

5. Indigenous groups: local peasants in various places are trying to win basic rights; examples are the Chiapas Zapatistas in Mexico, and activism in presidential elections in Bolivia, Ecuador, and elsewhere; Bolivians elected an Aymara Indian (Evo Morales) as its president in 2005

VI. Economic and Social Development: Dependent Economic Growth

A. Most Latin American countries are "middle income" according to World Bank categories, but extreme poverty also exists in the region
1. International debt has been a problem
2. Oil brings income into Mexico, Ecuador, Venezuela

B. Development Strategies
1. Extensive industrial and infrastructure development in the 1960s and 1970s, but led to debt and rural displacement
2. *Neo-liberal policies* ("Washington Consensus") (privatization) were implemented in the 1990s; as economic downturns came, several countries have adopted populism

3. *Import substitution:* policies that foster domestic industry by imposing inflated tariffs on all imported goods, thus substituting domestic goods for imported

4. Industrialization: most government policies emphasize manufacturing, usually in capitol cities or in planned *growth poles*: cities that, given a certain amount of investment, continue to attract more business, industry, and people, along with their investment; about 15–20 percent of workers in Mexico, Argentina, Brazil, Chile, Colombia, Peru, Uruguay, and Venezuela are in manufacturing

5. Maquiladoras and foreign investment: *Maquiladoras:* Mexican assembly plants (motor vehicles, consumer electronics, apparel) that line the border with United States; controversial because of the lower wages and weakened environmental regulations in Mexico; employment in maquiladoras peaked in 2001, when jobs in this area began moving to China

6. The entrenched *informal sector*: refers to the economic sector that relies on self-employed, low-wage jobs (e.g., street vending, shoe shining, artisan manufacturing) that are unregulated and untaxed; some people include illegal activities, such as drug activity or prostitution; squatter settlements are an example; because this sector is unregulated, no one really knows how large it is

C. *Primary Export Dependency:* dependence on the export of raw materials, including minerals and agricultural products (which are classified in the primary sector.

1. Agricultural production: mechanized and diversified (soybeans, rice, cotton, oranges, wheat, sugar, fruits, and vegetables) since 1960; some plantations survive; peasant farmers being squeezed out
2. Mining products: silver, zinc, copper, iron ore, bauxite, gold
3. Forestry (logging): mostly unsustainable; certification programs designate when a wood product has been produced sustainably; plantation forests becoming common and reduce pressure on forests of native species
4. The energy sector: Mexico, Venezuela, Ecuador export oil; Brazil becoming a world leader in ethanol based on sugar cane

D. Latin America in the Global Economy
1. *Dependency theory:* this theory says expansion of European capitalism created Latin America's condition of underdevelopment (1960s); these countries are vulnerable to changes in the global market; trade within Latin America provides another path to development
2. Neoliberalism as globalization: *neoliberal policies* stress privatization, export production, direct foreign investment, and few restrictions on imports; Chile has seen good growth using this strategy
3. *Dollarization:* the process by which a country adopts, either in whole or in part, the United States dollar as its official currency; Panama adopted this policy in 1904, Ecuador in 2000; this strategy (whether implemented in part or in full) helps to reduce inflation and fear of currency devaluation; countries that dollarize no longer control their own monetary policy

E. Social development: vast improvement in life expectancy, child survival, educational equality since 1960, partly as a result of grass-roots and nongovernmental organizations; regional differences exits within individual countries
1. Race and inequality: while there is relative tolerance for difference, Amerindians and blacks are overrepresented among the poor; class (and race) awareness is strong
2. Status of women: contradictory indicators; women work outside the home at nearly the same rate as in the United States, in most countries of Latin America, more women than men attend high school; women involved in politics, including a woman elected president of Nicaragua in 1990, and another in Panama in 1999; in 2005, Chile elected Dr. Michelle Bachelet as its president

VII. Conclusions

Latin America was the first region fully colonized by Europe, and as much as 90 percent of the population died as a result. The demographic recovery has been slow, but has included a rich diversity of people. This region is richer in natural resources than Asia, but there is concern that the region may be quickly exploited for short-term gain rather than for sustainable development. The region has an active informal economy, and rapid development. Within the region, and even within individual countries, there remain many disparities from one place another.

PRACTICE MULTIPLE CHOICE QUIZ

1. What are the boundaries of Latin America?
    a. Amazon rain forest in the north to Montevideo in the south
    b. Llanos in the north to Patagonia in the south
    c. Nicaragua in the north to Bahia Blanca in the south
    d. Rio Grande in the north to Tierra del Fuego in the south
    e. Yucatan Peninsula in the north to La Plata in the south

2. In several countries of Latin America, there is still a strong Indian presence. Which of the following countries does NOT have a strong Indian presence?
    a. Bolivia
    b. Ecuador
    c. Guatemala
    d. Peru
    e. Uruguay

3. Latin America has not experienced the same level of environmental degradation as North America, East Asia, and Europe. What is the reason?
    a. Relatively low population density
    b. Large size
    c. A long-enduring respect for the environment
    d. A and B above
    e. B and C above

4. Of all the environmental problems in Latin America, which one is the most critical for biological diversity?
    a. Air pollution
    b. Loss of tropical rain forest
    c. Smog
    d. Subsidence
    e. Water quality

5. What have been the results of the use of farm chemicals (pesticides, herbicides, fertilizers) in Latin America?
    a. Contamination of groundwater supply
    b. Rashes and burns on the skin of farm workers
    c. Rise in serious birth defects
    d. A and B above
    e. A, B, and C above

6. Which of the following statements about the Andes Mountains is FALSE?
   a. They are relatively young
   b. They are nearly 5000 miles long
   c. The Andes are the world's tallest mountains
   d. Volcanoes and earthquakes are common in the Andes
   e. Rich veins of precious metals and minerals are found in the Andes

7. What is the key element that causes altitudinal zonation?
   a. Wind caused by monsoons
   b. Variations in precipitation caused by orographic precipitation
   c. Temperature drop known as environmental lapse rate
   d. Storms caused by El Niño
   e. All of the above

8. What weather pattern signals the arrival of El Niño?
   a. Unseasonably cool temperatures
   b. Torrential rains
   c. Tornadoes
   d. Monsoonal winds
   e. Drought

9. What is the country of origin of the largest number of legal immigrants to the U.S.?
   a. Venezuela   b. Argentina   c. Nicaragua   d. Mexico   e. Panama

10. In which of the following countries of Latin America is there no urban primacy?
    a. Brazil   b. Guatemala   c. Mexico   d. Peru   e. Venezuela

11. What was the basis for political and economic power in Latin America since the colonial era?
    a. Military weapons
    b. Grass-roots organization
    c. Education
    d. Control of land
    e. All of the above

12. From which countries did migrants come to Latin America to work in Brazil and Peru?
    a. North and South Korea
    b. Libya and Egypt
    c. Indonesia and Malaysia
    d. India and Pakistan
    e. China and Japan

13. What is the key factor that unifies Latin America as a region?
    a. Climate
    b. Cultural unity
    c. History of colonization
    d. The Amazon Basin
    e. The Andes Mountains

14. Which country of Latin America was at the center of the drug trade when it began in the 1970s?
    a. Argentina   b. Brazil   c. Chile   d. Colombia   e. Mexico

15. What is a maquiladora?
    a. A squatter settlement in Latin America
    b. Mexican assembly plants that line the border with the U.S.
    c. A small plot of land provided to peasants in Latin America for their subsistence
    d. A person of European and Indian ancestry
    e. A cultural trait ascribed to women in Latin America

Answers: 1-D; 2-E; 3-D; 4-B; 5-E; 6-C; 7-C; 8-B; 9-D; 10-A; 11-D; 12-E; 13-C; 14-D; 15-B

Chapter 5
THE CARIBBEAN

LEARNING OBJECTIVES
- This chapter introduces the Caribbean, a region of great cultural diversity, which bears the imprint of its history of colonialization.
- This chapter provides the opportunity to compare and contrast two seemingly similar regions, Latin America and the Caribbean. By comparing and contrasting the two regions (which are very dissimilar), it will become apparent why the authors have chosen to separate them into different regions.
- The authors introduce the concept of plantation agriculture, which brings with it a number of physical and cultural changes, including mono-crop cultivation, seasonal labor (and seasonal unemployment), migration, and sometimes the introduction of a matriarchal society. Plantation agriculture has played an important role in other regions as well. Understanding the plantation system will be an aid to understanding other regions, too.
- Tourism is introduced here as an important economic activity.
- After studying this chapter, the student should be able to identify the countries of the Caribbean and locate specific physical features in the region.
- In addition, the student should understand the following concepts and models:

  - Plantation agriculture, "Plantation America"
  - Monocrop production
  - Hurricanes
  - African diaspora
  - Maroon societies
  - Free trade zones
  - Circular and chain migration
  - Monroe doctrine
  - Offshore banking

CHAPTER OUTLINE

I. Introduction

The Caribbean contains more than 41 million people in 26 countries and dependent territories, located predominantly in or on the Caribbean Sea. In addition to the many islands in the region, Belize (in Central America) and the three Guianas (Guyana, Suriname, and French Guiana) on the continent of South America are also included as part of the Caribbean region. Rival European powers battled each other for control of the region, but early in the 1900s, the United States became the dominant geopolitical force in the Caribbean. Agricultural activity (including plantations) remains important in this region, and population densities are high. Environmental problems abound. Caribbean countries are looking to tourism, offshore banking, manufacturing, and nontraditional exports (e.g., flowers) to reduce dependence on the exports of primary sector products (especially plantation crops). There are marked disparities in individual wealth in the countries of the Caribbean and among people within individual countries as well. This region illustrates the concept of *isolated proximity*, in which geographic isolation of the Caribbean sustains its cultural diversity and limited economic opportunities, while its proximity to North America provides it with transnational connections and economic dependence.

II. **Environmental Geography: Paradise Undone**
  A. Introduction: the story of Haiti
    1. Once France's richest colony, Haiti's per capita income is now just $480, and its life expectancy (58 years) is the shortest in the Americas
    2. Colonization, slavery, political corruption, unwise land stewardship, population pressure have all played a role
  B. Environmental Issues (Fig. 5.4)
    1. Environment in the Caribbean is vastly changed from original; severe depletion of biological resources (plants and animals)
    2. Agriculture's legacy of deforestation: caused by plantation agriculture (including sugar cane), and the need for fuel and lumber; deforestation causes soil erosion, loss of biodiversity
    3. Managing the rimland forests and coasts: biological diversity in this region is better protected than on the islands; but international logging and economic development cause concern

  4. Urban environmental problems: water contamination and waste disposal damage environment; water quality and quantity are problems, especially for the urban poor; only half of Haiti's people have access to clean drinking water
  C. The Caribbean and climate change
    1. Bahamas could lose 30% of land with 10 ft rise in sea level; other islands affected.
    2. Region contributes little greenhouse gas (GHG), is very vulnerable to climate change, especially coral reefs
    3. CARICOM monitors climate change threats
  D. Islands and Rimland Landscapes
    1. Connected through trade routes; the sea has biological diversity, but there is not enough of any one species for commercial fishing; includes the Antillean Islands (Greater and Lesser Antilles)
    2. There is growing recognition of the economic value of clean environment
    3. General physical geography (Fig. 5.5); region lies between Tropic of Cancer and the equator; warm temperatures (70s year-round); diverse flora and fauna
    4. Greater Antilles: Cuba, Jamaica, Puerto Rico, Hispaniola (Haiti and Dominican Republic); home to 86% of the Caribbean's people; best farmland, but many soils leached; mountains range from 6000–10,000 feet
    5. Lesser Antilles: small islands from Virgin Islands to Trinidad; smaller and with fewer people than Greater Antilles; volcanic origins, Montserrat volcano

active since 1995; agriculture on islands of Barbados, Antigua, Barbuda, eastern half of Guadaloupe

6. Rimland States: *rimland:* coastal zone of the mainland, beginning with Belize and extending along the coast of Central America to northern South America; Belize and Guianas are rimland states in the Caribbean; Belize is limestone, Guianas have rolling hills; rimland retains much of its forests

E. Climate and Vegetation (Fig. 5.7)
1. Warm year-round
2. Abundant rainfall (>80 inches), producing forests and grasslands
3. Dry basin in western Hispaniola
4. *Hurricanes:* storms with heavy rains and fierce winds (75 mph or higher); hurricane zone lies just north of equator; better prediction technology helps prevent loss of life (evacuations); but property damage is great
5. Forests: much of the original forest on islands cleared for Europeans' plantations; today most land used for export crops (e.g., sugar cane); the Guianas' rain forests are mostly intact, because economies rely on mining, but pressure on mahogany forests increases
6. Savannas and mangroves: Hispaniola and Cuba's palm savannas have fertile grasslands, are good farmlands; mangrove swamps are a vital nursery for young crustaceans and fish; removing mangroves promotes erosion

III. **Population and Settlement: Densely Settled Islands and Rimland Frontiers**
A. Generally High Population Densities
1. 86% of Caribbean's people live on the Greater Antilles
2. Puerto Rico has highest population density
3. Mainland Belize and Guianas more lightly populated
B. Demographic Trends
1. Early reliance on importation of enslaved Africans to balance high mortality among indigenous peoples under European colonization
2. Natural population increase began in mid- to late 19th century (1800s)
3. Fertility decline: Cuba, and others have TFR under 2.0
4. The rise of HIV/AIDS: important regional issue; 1.6% of Haitians ages 15–49 infected; tourism and prostitution may play a role in transmission of the disease; infection rates have dropped because of education
5. Emigration: Caribbean people began emigrating to other Caribbean Islands, North America, Europe in 1950s, mostly for jobs and economic opportunity (push factors), contributing to a *Caribbean diaspora:* the economic flight of Caribbean peoples across the globe (Fig. 5.12)
6. *Circular migration:* when a man or woman leaves children behind with relatives in order to work hard, save money, and return home; *chain migration:* when one family member at a time is brought to the destination.

C. The Rural-Urban Continuum
1. Structure of Caribbean communities reflects plantation legacy; many subsistence farmers are descendents of former slaves; rural communities are loosely organized, and there is a pattern of temporary, seasonal migration of men looking for farm work, resulting in matriarchal family structure
2. Caribbean cities: surge in rural-urban migration since 1960s because of fewer rural jobs, now 64 percent of people in Caribbean live in cities; city morphology often similar to those in Latin America because of shared Spanish influence; but Paramaribo, Suriname resembles a "tulipless Holland" because of its colonization by Dutch; British, French cities not distinctive
3. Housing: shortage of urban opportunities and housing has resulted in squatter settlements and reliance on informal sector; Cuba is an exception, its apartments reflecting the influence of Russia and socialism

IV. **Cultural Coherence and Diversity: A Neo-Africa in the Americas**
A. Common historical and cultural processes unite the region, whose people have many linguistic, religious, and ethnic differences; *Creolization:* blending of African and European cultures in the Caribbean
B. The Cultural Imprint of Colonialism
1. European colonies destroyed indigenous society and imposed different cultural and social systems; demographic collapse as in Latin America; after collapse, Europe brought enslaved Africans to work in the region
2. *Plantation America:* a culture region that extends midway up the coast of Brazil, through the Guianas and the Caribbean, and into the southeastern United States (phrase coined by Charles Wagley); the term describes a production system that endangered ecological, social, and economic relations; the plantation system gave elites control of the land, caused rigid class lines and the formation of a multiracial society that favored lighter-skinned people
3. *Mono-crop production:* an agricultural system that concentrates on one single crop, usually a commodity such as sugar or bananas, characteristic of plantations
4. Asian immigration: most Asian immigrants to the Caribbean came as *indentured laborers:* workers contracted to work on estates for a set period of time (usually several years), from South and Southeast Asia after slavery ended in mid-19th century (mostly from India, Indonesia, some from China); this trend most prominent in Guyana, Trinidad, Suriname
C. Creating a Neo-Africa: enslaved Africans introduced in 16th century, after demographic collapse
1. *African diaspora:* the forced removal of people from their native lands; slave trade crossed Sahara, included North Africa, and linked East Africa with Middle East slave trade (see

45

Chapter 6); 12 million-plus enslaved Africans landed in the Americas, half landing in the Caribbean; about 2 million more died en route (Fig. 5.16)

2. Maroon societies: *Maroons:* runaway slaves who formed their own communities; surviving Maroon settlements protected African traditions (farming practices, house designs, community organization, language, religion); examples: Palmeros (in Pernambuco, Brazil), existed throughout 17th Century; several remain today (Fig. 5.17)

3. African Religions; linked to maroon societies, but evolved into unique forms with ties to West Africa; examples: Voodoo (Vodoun) in Haiti, Santeria in Cuba, Obeah in Jamaica (Fig. 5.18)

D. Creolization and Caribbean Identity

1. *Creolization:* the blending of African, European, and even some Amerindian cultural elements into the unique sociocultural systems found in the Caribbean

2. Reflects influences of the many peoples of the region

3. Especially recognizable in language and music

E. Language (Fig. 5.20)

1. European languages (Spanish, French, English, Dutch) dominate; Dutch declining, French Creole (also called *patois,* with African syntax) official in Haiti; reflects and instills a sense of identity

2. Music: rhythmic styles (reggae, calypso, merengue, rumba, etc.); steel pan drums of Trinidad; linked with Afro-Caribbean religions; may have political overtones (like Bob Marley's tunes)

V. **Geopolitical Framework: Colonialism, Neo-Colonialism and Independence**

A. Europeans viewed the Caribbean as a strategic and profitable region

B. United States' presence in Caribbean increased after adoption of *Monroe Doctrine,* which claimed that the United States would not tolerate European military involvement in the western hemisphere

C. Life in the "American Backyard"

1. The Caribbean has been described as the "American Backyard"

2. United States built Panama Canal, eventually returned the Canal Zone to Panama

3. Other policies cast United States as imperialist: Good Neighbor Policy (1930s); Alliance for Progress (1960s); Caribbean Basin Initiative (1980s)

4. Imperialism requires the ability to impose one's will, by force if necessary; United States has had embargoes against various Caribbean countries; United States occupied Dominican Republic (1916–24), Haiti (1913–34), Cuba (1906–09, 1917–22) and sent troops to Haiti in 1994 (Fig. 5.23)

D. The Commonwealth of Puerto Rico

1. Puerto Rico is a commonwealth (possession) of the United States

2. Independence movements seek secession from United States

3. Operation Bootstrap: Industrialization program with tax incentives and low-cost labor

lured United States textile, petrochemical, and pharmaceutical firms to Puerto Rico

E. Cuba and Regional Politics
1. Cuba was an ally of the former U.S.S.R. (Russia)
2. Began as a Spanish colony, gained freedom in 1895–98 war (United States joined at the end)
3. Revolution of 1950s: although Cuba's economic development was good, there was a large gap between rich and poor; in 1959, Cubans overthrew their government, and put Fidel Castro in power of a socialist system; it still exists, although Castro stepped down in February 2008
4. Nationalized U.S. industries in Cuba, and took ownership of all foreign-owned property
5. Under Castro, Cuba's literacy and public health have improved
6. Castro established a formal relationship with the U.S.S.R. in 1960; the Cuban Missile Crisis happened in 1962

F. Independence and Integration
1. Independence movements: most countries are independent
2. Present-day colonies: Britain: Cayman Islands, Turks and Caicos, Montserrat; France: French Guiana, Martinique, Guadeloupe; Federation of the Netherlands Antilles: Curacao, Bonaire, St. Martin, Saba, St. Eustatius
3. Regional integration: Caribbean Community and Common Market (CARICOM), a regional trade association (1972) is trying to integrate the region's countries economically

VI. Economic and Social Development: From Cane Fields to Cruise Ships
A. From Fields to Factories and Resorts: the countries of the Caribbean are classified as lower middle income
1. Agriculture is in decline as a source of economic development
2. Sugar: sugarcane was an important innovation, now less important because of substitutes; a plantation crop
3. Coffee: coffee beans are grown in the mountains of the Greater Antilles, mostly by small farmers who also grow subsistence crops. The instability of coffee prices is a problem.
4. The banana wars: Latin America grows more bananas than the Caribbean, and Latin American bananas are the large, yellow, unblemished Cavendish variety that the world market prefers. In the Caribbean, export of bananas has fostered economic and social development. Increased global pressures are making it difficult for Caribbean banana growers to compete.

B. Assembly-Plant Industrialization
1. Foreign companies invited to build factories in the Caribbean (like Puerto Rico's Operation Bootstrap of 1950s)
2. Creation of *free trade zones (FTZs):* duty-free and tax-exempt industrial parks for foreign corporations, which legalize foreign ownership and provide cheap labor; these include assembly plants in major cities of the Caribbean
3. Depends on national and international policies that support

export-led growth through foreign direct investment

C. Offshore Banking
1. *Offshore banking centers* appeal to foreign banks and corporations by offering specialized services that are confidential and tax-exempt; cities with these centers make money from registration fees, not taxes; Bahamas were third largest banking center in the world in 1983; now Cayman Islands are the leader; proximity to United States is appealing
2. These centers also attract money tied to drug trade, terrorist groups (money laundering)
3. Offshore banking has relatively few employees
4. Online gambling has been growing in the region since 1999, when the WTO allowed it

D. Tourism
1. Cuba's earlier leading role in tourism stopped with rise of Castro
2. Other islands (Puerto Rico, Bahamas, Jamaica, Dominican Republic) now popular. Cuba is also popular with visitors from outside the United States
3. Problems with tourism: subject to overall health of world economy, since travel is a luxury good; vulnerable to natural disasters; *capital leakage:* the huge gap between gross receipts and the total tourist dollars that remain in the Caribbean (multinational firms own many resorts)
4. Advantages: promotes stronger environmental laws; environmentally less destructive than export agriculture; now more profitable than export agriculture

E. Social Development
1. Measures of social development are improving, although Haiti is in very bad shape
2. Education: Cuba and the English Caribbean (Bahamas) excel, with high literacy and secondary school graduation rates; Haiti and Dominican Republic (Hispaniola) have 20 percent and 30 percent secondary school graduation rates; *brain drain:* a large percentage of the best-educated people leave the region, usually to more developed places, thus providing a subsidy to the developed regions
3. Status of women: matriarchy important, as many men leave home for seasonal work; woman control many activities, but do not have high status; women often are excluded from the cash economy; more women are working in assembly plants, tourism; greater access to cash challenges existing gender roles; smaller families are resulting
4. Labor-related migration: Intra-regional, seasonal migration is traditional; after WWII, many have migrated to North America
5. *Remittance:* money that migrants send back home; collectively, this can add much to a country's economy; some international migrants return home to retire, bringing cash, ideas, and positive economic and political changes

VII. Conclusions
The Caribbean is better integrated into the global economy than most of the developing world. Its European influence is still

apparent in its economic and urban systems. Although agriculture was an important part of the region's economic development, today industrialization, banking, and tourism are the major sources of development.

PRACTICE MULTIPLE CHOICE QUIZ

1. When Europeans began to colonize the Americas, where did they start?
    a. Latin America
    b. The Caribbean
    c. The east coast of what is now Canada
    d. The west coast of what is now the United States
    e. Yucatan Peninsula

2. In which part of the Caribbean would you find the majority of the region's population, arable lands, and large mountain ranges?
    a. Bahamas
    b. Greater Antilles
    c. Lesser Antilles
    d. The Rimland
    e. Virgin Islands

3. What feature defines seasons in the Caribbean?
    a. Temperature variation
    b. Shifting wind patterns
    c. Changes in rainfall
    d. A and B above
    e. A, B, and C above

4. All of the following statements about hurricanes in the Caribbean are true EXCEPT....
    a. Caribbean hurricanes originate off the west coast of Africa
    b. Hurricanes typically enter the Caribbean through the Lesser Antilles
    c. The hurricane season lasts from July to October
    d. Winds of 75 mph or higher are required for designation as a hurricane
    e. The average hurricane season produces 20 or more hurricanes

5. Today, where are the tropical forests most commonly found in the Caribbean?
    a. Bahamas
    b. Greater Antilles
    c. Lesser Antilles
    d. The Rimland
    e. Virgin Islands

6.	What is the most significant demographic trend in the Caribbean?
	a.	Decline in fertility
	b.	Drop in population
	c.	Increase in family size (TFR) throughout the region
	d.	Increased immigration from Russia since the breakup of the Soviet Union
	e.	Increasing rate of natural increase (RNI)

7.	Which of the following matchups between Caribbean migrants and their destinations is INCORRECT?
	a.	Barbados residents go to England
	b.	Cubans go to Russia
	c.	Haitians go to the Dominican Republic
	d.	Jamaicans go to North America
	e.	Surinamese go to the Netherlands

8.	Caribbean cities in areas colonized by the Spanish tend to look like those in which other world region?
	a.	Latin America
	b.	Sub-Saharan Africa
	c.	South Asia
	d.	North America
	e.	Europe

9.	In which country of the Caribbean are squatter settlements LEAST common?
	a.	Bahamas	b. Cuba	c. Dominican Republic	d. Haiti	e. Jamaica

10.	Why were indentured workers from South Asia needed in the Caribbean?
	a.	Illness killed most of the slaves and peasant farmers
	b.	Slaves had been freed, and slavery made illegal, eliminating that labor source
	c.	The colonizers found additional markets and needed more workers for their expanded production
	d.	The work became harder as lands became less productive
	e.	All of the above

11.	Where in the Caribbean did Maroons tend to settle?
	a.	Along the rivers?
	b.	In the cities
	c.	On the best farmlands
	d.	On the coasts
	e.	In isolated areas

12. Why were plantations described as "monocrop?"
    a. Because of the number of markets for their products
    b. Because of their ranking in terms of agricultural productivity, compared to other types of agricultural
    c. Because the most common number of commodities grown their was 1 (one)
    d. Because they typically had just one manager
    e. Because they were the number one favorite type of agriculture in the region

13. Which part of the Caribbean has the U.S. NOT invaded?
    a. Bahamas   b. Cuba   c. Dominican Republic   d. Grenada   e. Haiti

14. What makes Cuba distinctive among Caribbean countries?
    a. It has the highest elevation
    b. It has the highest percentage of rainforests remaining
    c. It has the smallest population in the region
    d. It is communist
    e. It is the only monarchy in the region

15. Which of the following is not a current economic activity in the Caribbean?
    a. Computer software development
    b. Banking
    c. Assembly-plant industrialization
    d. Sugarcane and coffee cultivation
    e. Tourism

Answers: 1-B; 2-B; 3-C; 4-E; 5-D; 6-A; 7-B; 8-A; 9-B; 10-B; 11-E; 12-C; 13-A; 14-D; 15-A

Chapter 6
SUB-SAHARAN AFRICA

LEARNING OBJECTIVES

- This chapter introduces Africa south of the Sahara Desert. This is a region that was in the center of Gondwanaland, and is the birthplace of humankind.
- The trade of enslaved Africans is an important aspect of this region's history, and one the student should understand.
- The student should also be aware of the key role that Europeans played in drawing the boundaries of states in this region, as well as other consequences of colonization of the region.
- Students should also understand the role that diseases have played in this region.
- At the conclusion of this chapter, the student should be familiar with the physical, demographic, cultural, political, and economic characteristics of Africa.
- In addition, the student should understand the following concepts and models:

  · Gondwanaland
  · Desertification
  · Agricultural density
  · Swidden (shifting cultivation)
  · Trade of enslaved Africans
  · Apartheid
  · Berlin Conference
  · Structural adjustment programs

CHAPTER OUTLINE
I. Introduction

Sub-Saharan Africa includes 48 states and one territory. The region has more than 7800 million people and is the world's fastest growing region (2.5 percent RNI). This region has seen incredible impacts from colonial powers, including the enslavement and forced removal of millions of its residents over several centuries and the political partitioning of the land by Europeans in 1884. The coherence of this region is based largely on the similarity of livelihood systems and the shared colonial experience among the peoples and states of this region. This region is culturally complex; in the large countries as many as 20 different languages may be spoken–most Africans speak several languages. African culture and musical influences have diffused around the world, as the African diaspora spread. The economy of Africa trails those of Latin America and the Caribbean. World Bank *structural adjustment programs* attempt to reduce government spending and encourage private sector initiatives; often they trigger drastic cutbacks in government-supported services and food subsidies, hurting the poor disproportionately hard. Philanthropic organizations and small, indigenous political and social movements are helping to meet basic needs and empowering communities to solve their own problems in their own ways. With the significant resources available in this region, there is reason to believe that Africa will make significant progress in the coming years.

II. Environmental Geography: The Plateau Continent

A. General description (Fig. 6.1) Africa is the largest landmass straddling the Equator, beautiful terrain but relatively poor soils

B. Plateaus and Basins
   1. Plateaus, elevated basins, dominate interior Africa; Africa is an elevated landmass, but its mountains are not relatively low; in the Rift Valley, volcanic mountains Kilimanjaro (19,000 ft) and Mount Kenya (17,000 ft), are Africa's two tallest peaks
   2. *Great Escarpment:* rims Southern Africa, beginning in southwest Angola, ending in northeast South Africa at the Drakensburg Mountains; it has been an impediment to coastal settlement
   3. Watersheds (four major rivers): (1) Congo is largest in the region, second only to Amazon in discharge to ocean; rapids and falls limit navigability; (2) Nile is world's longest river; begins in lakes (Victoria and Edward) in Rift Valley highlands; important water source for Sudan; (3) Niger is critical source of water from Mali and Niger; originates in Guinea highlands; (4) Zambezi is much smaller, begins in Angola and flows east; has Africa's two largest hydroelectric plants
   4. Soils: most fertile soils in Rift Valley (volcanic origin), Lake Victoria lowlands, central highlands of Kenya; elsewhere, soil is relatively infertile because much organic material and nutrients have leached away

C. Climate and Vegetation (Fig. 6.13)
   1. Warm year-round (70–80 degrees)
   2. Variable rainfall differentiates regions: high of 50 inches in Addis Ababa, Ethiopia; less than 1 inch in Namibian Desert
   3. Mountain zones exhibit altitudinal (vertical) zonation

D. Three biomes: tropical forests, savannas, and deserts
   1. Tropical forests: second-largest expanse of tropical rain forest in Congo Basin; year-round precipitation; warm-to-hot temperatures; a greater percentage remains than in Latin America or Southeast Asia; low population and subsistence agriculture may explain this; in the future, increasing international demand for these forests may cause greater pressure to cut
   2. Savannas: surround Central African rain forest belt; wetter savannas next to forests have trees and tall grasses; in drier zones, there are fewer trees and shorter grasses; wet savannas with forests south of the Equator provide habitat for large mammals (elephants, zebras, rhinos, lions)
   3. Deserts: Sahara means desert; Sahara is world's largest desert, one of the driest, spans Africa from Atlantic coast to the *Horn of Africa* (northeast corner that includes Somalia, Ethiopia, Djibouti, Eritrea); Namib Desert on west coast of Namibia has mild temperatures, and rainfall is very rare; Kalahari desert not quite dry enough to be a true desert, with just over 10 inches/year, but surface water is rare

E. Africa's Environmental Issues (Fig. 6.9)
   1. Close links between land and people mean environmental shortages are readily felt
   2. The Sahel and desertification: the *Sahel* is the semi-desert region south of the Sahara; *desertification* is the expansion of desert-like conditions as a result of human degradation; causes include (1) improper cultivation; (2) cultivation of non-native species by colonialists; (3) overgrazing the land; and (4) global atmospheric cycles that result in cyclical droughts; some people believe these lands have too many people and that desertification will always be a problem
   3. *Transhumance:* the movement of domesticated animals between wet-season and dry-season pastures
   4. Drought and famine: drought is a recurring problem in Africa and often results in food shortages and famine. Other causes of famine include weather, misrule, war, disease, poverty.
   5. *Deforestation:* north of the Equator, forests are being converted to grasslands, savanna, and sometimes land for crops; *biofuels* (wood and charcoal used for household needs, especially cooking) are now in short supply; local groups are replacing trees, with Kenya's Green Belt Movement (most of whose participants are women) one of the most successful
      Wildlife conservation: Africa has some of the world's most impressive large mammals; much of these animals' habitat has been destroyed because of (1) population pressure, (2) political instability, (3) poverty, (4) poaching (especially rhinos), and (5) the ivory trade; the most secure wildlife preserves are in southern Africa and are tourist destinations

F. Climate change and vulnerability in Sub-Saharan Africa
   1. This region is the lowest emitter of GHG, is very vulnerable to climate change
   2. Some land may become more productive, some less productive
   3. Possible increase in vector-borne diseases
   4. Famine could worsen

III. Population and Settlement: Young and Restless
   A. Quick Growth: it is estimated that the population in this region will grow by 128% by 2050
      1. Young population (44 percent under age 15; compared to 18 percent in developed countries); large families (5–6 children); high maternal/infant mortality
      2. Relatively low density (similar to United States)
      3. *Physiological density:* number of people per arable (farmable) land
      4. *Agricultural density:* ratio between the number of farmers per unit of arable land; measures number of people who directly depend on arable land
   B. Population Trends and Demographic Debates
      1. Has Africa reached its carrying capacity?
      2. Family size: cultural practices, rural lifestyles, economic realities promote large families (5–6 children); reasons for large families: (1) women marry young, often have little formal education; (2) ethnic rivalries may encourage pronatal policies; (3) high infant mortality encourages many children; and

54

(4) large families an asset in rural areas; increasing urbanization and government policies promote smaller families
    3. The impact of AIDS on Africa (Fig. 6.16): 2/3 of AIDS cases are in Sub-Saharan Africa; infection rates high enough to slow population growth; advanced medications too costly for most people; education helping stop spread of HIV/AIDS
C. Patterns of Settlement and Land Use (Fig. 6.18)
    1. Most Africans live in rural areas, some cities becoming larger, the city of Lagos, Nigeria, has about 14 million people (it's a megacity); West Africa more heavily populated than the rest of Africa; population widely scattered,
    2. Agricultural subsistence: *swidden* (shifting cultivation or slash-and-burn) common; Madagascar's irrigated rice fields introduced by Indonesian settlers 1,500 years ago; shifting cultivation was introduced by migrants from Africa later
    3. Plantation Agriculture: export crops important to many African economies; coffee (Ethiopia, Kenya, Rwanda, Burundi, Tanzania); peanuts (Sahel); cotton (Sudan, Central African Republic); cocoa (Ghana, Ivory Coast); plantation rubber (Liberia); palm oil (Nigeria)
    4. Herding and livestock: especially important in semi-arid regions; *pastoralists:* people who specialize in raising grazing animals; camels and goats in and south of the Sahara, cattle farther south; *tsetse flies* spread sleeping

sickness to cattle and people, make much of Sub-Saharan Africa unacceptable for cattle herds
D. Urban Life
    1. Sub-Saharan Africa is the least urbanized region in developing world, but urbanization is increasing; there's a tendency toward urban primacy (e.g., Lagos, Nigeria)
    2. East African urban traditions: ancient cities like Axum (Ethiopia) are 2,000 years old; Tombouctou, and Gao more than 1000 years old; urban mercantile culture rooted in Islam and Swahili language; trade in Zanzibar (Tanzania) and Mombasa (Kenya)
    3. West African urban traditions: most cities in West Africa combine Islamic, European, national elements; Yoruba cities in southwestern Nigeria have walls, gates, palaces
    4. Urban industrial South Africa: most major cities in southern Africa have colonial origins; South Africa most urbanized country in Africa, industries based on diamonds, gold, chromium, platinum, tin, uranium, coal, iron ore, manganese; cities in South Africa racially segregated (Fig. 6.23)
    5. Apartheid was the official policy of racial segregation in South Africa; whites, blacks, coloreds (mixed African and European ancestry), Indian were the races

IV. **Cultural Coherence and Diversity: Unity through Adversity**
    A. Language Patterns (Fig. 6.25)

1. Complex pattern with local languages, African trade languages, and languages from Europe and Asia
2. African language groups: Niger-Congo most important in the region, includes Mandingo, Yoruba, Fulani, Igbo, Swahili (which is important to trade); Nilo-Saharan (Sudan, Sahel, East African); Khoisan (Kalahari)
3. Language and identity: ethnic identity in Africa has been fluid, but Europeans artificially divided some groups, misinterpreted other territorial boundaries, and applied meaningless names to groups; *tribes:* a group of families or clans with a common kinship, language, definable territory; in most African countries, multiple languages are the norm
4. European languages (Fig. 6.25): colonial powers promoted use of their languages, especially in government and education; most Sub-Saharan African countries continued this use after colonizers left; Francophone Africa is French-speaking, Anglophone Africa speaks English, Afrikaans (Dutch-based) in South Africa; Arabic in Mauritania and Sudan

B. Religion
1. Indigenous African religions generally classed as *animist:* centered on the worship of nature and ancestral spirits; animist religions vary greatly
2. Introduction and spread of Christianity: brought to northeast Africa around A.D. 300; half of Ethiopians and Eritreans are Coptic Christians; Protestantism began arriving in 1600s; Catholicism followed, especially in French, Belgian, and Portuguese colonies; Jewish community in Johannesburg; current missionaries in Africa include Pentecostals, Evangelicals, Mormons
3. Introduction and spread of Islam (Fig. 6.9 began around 1,000 years ago; Senegal was the first Sub-Saharan Muslim state; Islam prevails in most of the Sahel today
4. Interaction between religious traditions: religious conflict most acute in northeast Africa between Muslims and Christians; Islam and Christianity continue to spread, as animism remains

C. Globalization and African Culture
1. Role of slavery (Fig. 6.31): undermined demographic and social strength of African societies; estimated 12 million Africans enslaved from 1500–1870; most Africans sent to plantations in the Americas, but many taken to Europe, North Africa
2. Popular culture in Africa mixes local and global influences; Lagos is Africa's film capital; "Ethiopian Idols" is like its United States and British counterparts
3. Music in West Africa: Nigeria and Mali lead popular music in the leader; singer Fela Kuti became a voice of political conscience for Nigerians struggling for democracy; lyrics criticized military government; others copied his style, while the state harassed him

4. The pride of runners: Ethiopia and Kenya, have produced many world-class distance runners

V. Geopolitical Framework: Legacies of Colonialism and Conflict

A. Africa Is the birthplace of humanity
B. Indigenous Kingdoms and European Encounters
   1. Nubia was the first significant state of Sub-Saharan Africa, founded in northern Sudan about 3,000 years ago, using political models from Egypt and Arabia; Axum arose in northern Ethiopia and Eritrea 2,000 years ago (Egypt/Arab political model); other states followed
   2. Early European encounters: some African states increased their military and economic power by participating in the slave trade, often using money from selling enslaved peoples to buy weapons; it took Europeans many centuries to gain effective control of Sub-Saharan Africa
   3. The Disease Factor: malaria and other tropical diseases killed many Europeans, slowing colonization, but the prospect of African riches caused Europeans to persist; discovery of quinine to protect against malaria paved the way to European colonization
C. European colonization: some African states increased their military and economic power by participating in the slave trade, often using money from selling enslaved peoples to buy weapons; it took Europeans many centuries to gain effective control of Sub-Saharan Africa
   1. The Berlin Conference (Fig. 6.35): 13 European countries divided Africa; Africans left out entirely; lines the Europeans drew ignored existing African-drawn boundaries; Europeans' stronger weapons subdued Africans; Sudan last to be colonized in early 1900s; Ethiopia the only country to escape colonization by repelling the Italians; although Germany was a principal instigator of the scramble, it lost its colonies after World War I
   2. Establishment of South Africa: one of the oldest colonies of Sub-Saharan Africa and the first to gain freedom (1910); British settlers pushed out Dutch settlers (called Afrikaners or Boers); war between the two groups ensued; British won; in 1948, Afrikaners' National Party gained control of South Africa's government and instituted *apartheid* (formalized, systematic racial separation and discrimination); elements of apartheid: (1) petite—separate service entrances based on skin color, like Jim Crow in the United States; (2) meso—divided city into residential sectors by race (whites had best and largest neighborhoods); (3) grand—construction of black *homelands* (nominally independent lands that were rural, overcrowded, and on marginal lands, like United States Native American reservations); 3 million black people forcibly relocated to these areas, residence and even travel outside the homelands strictly regulated; blacks not allowed to vote
D. Decolonization and Independence
   1. Decolonization happened rather quickly and peacefully beginning

57

in 1957; some early independence efforts, like the Pan-African Movement, started in early 1900s; work of important African leaders set the stage for the founding of the Organization of African Unity (OAU), a continent-wide organization that tries to mediate disputes between neighboring African states

2. African Unity to African Union (AU): OAU was renamed African Union (AU) in 2002. AU is continent-wide, HQ in Ethiopia. AU initiated continental development plan: New Partnership for Africa's Development (NEPAD) promotes economic development, regional security, government reform and accountability

3. Apartheid's Demise in South Africa: opposition to apartheid began in 1960s with internal pressure from blacks, coloreds and Asians; external from corporate, political, and athletic ostracism; Nelson Mandela (a key black South African anti-apartheid leader) in 1990 after 27 years as a political prisoner; apartheid officially abolished in 1992; nonracial (all-race) elections in 1994, with Nelson Mandela elected; overall progress is slow, especially economic progress for nonwhites

E. Enduring Political Conflict (Fig. 6.40)
  1. The tyranny of the map: the boundaries drawn at the Berlin Conference have made it difficult to build cohesive states in Africa; Organization of African Unity

(OAU), founded in 1963, agreed to accept those borders

2. *Tribalism:* loyalty to the ethnic group rather than the state has made it difficult to unify African countries

3. *Refugees:* people who flee their state because of fear of persecution based on race, ethnicity, religion, or political beliefs

4. *Internally displaced persons:* people who have fled from conflict, but still remain in their country of origin

5. Ethnic Conflicts
  a. Genocide in Darfur, Sudan: Arab nomads *(janjaweed)* in concert with the government, targeted black sedentary farmers, African Union and the UN are trying to keep the peace; Conflict in the Congo from 1998-2004, now there's a fragile peace;
  b. Hutu vs. Tutsi in Rwanda, fueled by Belgian colonists who privileged the Tutsis, resulted in many killed in 1994; war crimes trials now underway
  c. Liberia, Sierra Leone endured civil war
  d. Somalia: clan warfare *(clan:* social units that are branches of a tribe or ethnic group larger than a family)

6. Secessionist movements: Eritrea seceded from Ethiopia (the only territory in the region to achieve this goal)

7. Big man politics: occurs when presidents (military or civilian) take the reigns of power and refuse to let go; since the 1990s, real democracy becoming more

common, and leaders are
stepping down when they are
voted out of office

VI. Economic and Social Development: The Struggle to Rebuild

A. Africa is the poorest, least-developed world region
  1. Region's economic base declined during the 1990s
  2. High population growth (2.5 percent in 2001)
  3. Turnaround possible: slow growth since 1995
B. Roots of African Poverty
  1. Traditional explanations: environmental problems, but these can be resolved; depopulation of many regions because of slavery, while other people fled to remote areas to escape enslavement; lack of investment in infrastructure, education, public health by colonists
  2. Failed development policies: several reasons for difficulties include decline in prices of commodities and products that come from Africa; policies of *economic nationalism* resulted in ventures in heavy industries (e.g., steel) in which they were not competitive, and maintenance of local currency at artificially high levels benefited local elites but made it difficult for Africans to sell their products, which were overpriced on the world market
  3. Food policies: governments kept crop prices too low; farmers unable to make profit; many shifted to subsistence and export crops, leading to food shortages
  4. Corruption: including bribery of government employees, and *kleptocracy* (a situation where corruption is so institutionalized that politicians and government bureaucrats skim a huge percentage of the country's wealth)
C. Links to the World Economy: connections with the world are limited; most exports go to the European Union, especially England and France, then the United States; connectivity infrastructure is inadequate; there is no African trade bloc (like Mercosur)
  1. Aid versus investment: Africa is linked globally more by aid than by the flow of goods or foreign direct investment; poverty and political instability discourage investment; debt relief is needed
  2. Debt relief: the World Bank and International Monetary Fund (IMF) propose to reduce debt levels in heavily indebted countries (such as in Africa); interest payments on the debt are very high
  3. China's growing role in the region: China-Africa forum in Beijing in 2006; China is investing in Africa in exchange for oil and other strategic raw materials; some have expressed concern about China's lack of consideration for human rights
D. Economic Differentiation within Africa
  1. There are great differences in economic and social development among African countries
  2. Life for the poorest: earning $12–$18 per month
  3. South Africa has a well-developed, well-balanced industrial economy; is a mining

superpower, but blacks are not as well-off as whites

4. Oil and mineral producers: Gabon, Namibia, Botswana have substantial oil and mineral reserves and small populations, leading to relatively good economic outcomes; Republic of Congo and Cameroon have oil, but have had economic declines in recent years

5. Leaders of ECOWAS: Nigeria has largest oil reserves in Sub-Saharan Africa, but high population keeps GNP low; Ivory Coast and Senegal still function as commercial centers, but have had economic downturns; Ivory Coast and Ghana economies are recovering

6. East Africa: Kenya is the commercial and communications center of East Africa, has good infrastructure, wildlife tourism, family size is declining, health improving; Uganda is improving and now has $1500 GNI/PPP

7. The poorest states are in the Sahel, the Horn, and the southeast; Mali, Burkina Faso, Niger, Chad have suffered long droughts and environmental degradation; Sudan, Ethiopia, Eritrea, Somalia, Mozambique, have suffered drought and war

E. Measuring Social Development
1. Actual figures reflect low level of social development, but trends are mostly positive, with child survival and literacy improving; access to education varies widely

2. Life expectancy in Africa is world's lowest (49 years): causes: (1) warfare and political disorder, (2) poor sanitation and infrastructure, (3) AIDS

epidemic, (4) high infant and child mortality rates (about 1 out of 5 children dies before reaching age five)

3. Health issues: doctors and health facilities are scarce; endemic diseases are severe and persistent (malaria, schistosomiasis, river blindness)

F. Women and Development: women's contribution to economy is mostly in informal sector and therefore invisible; women account for 75 percent of the labor that produces more than half the food consumed in the region

1. Status of women is mixed overall; they have considerable political and economic power, but damaging practices persist, including (1) polygamy, (2)"bride price," (3) denial of property inheritance, (4) female circumcision (genital mutilation)

2. Building from within: women's market associations, efforts to loan money to women to start small businesses, and investment in women's activities have paid off; for example, funding of women to plant trees has prevented soil erosion and enhanced future fuel supplies

VII. Conclusions

Sub-Saharan Africa has experienced many trials since its states gained independence. Political upheaval and environmental challenges have eroded economic gains of earlier times. Health risks (including AIDS) have also created a sense of pessimism about the prospects for improvement in the foreseeable future. In contrast, the relatively low population densities of Sub-Saharan Africa and the resource base do offer some reason for optimism.

PRACTICE MULTIPLE CHOICE QUIZ

1. What global landmark passes through Sub-Saharan Africa?
    a. Equator
    b. International dateline
    c. Arctic Circle
    d. A and B above
    e. A and C above

2. What makes Sub-Saharan Africa important from an anthropological perspective?
    a. Domestication of plants and animals occurred in this region first
    b. Human origins are traceable to this region
    c. It has the largest population of any region in the world
    d. People in this region speak the highest number of languages of any world region
    e. This is the source region of Christianity

3. What is the major environmental issue in Sub-Saharan Africa, and where is it most serious?
    a. Desertification in the Sahel
    b. Drought in the Congo
    c. Flooding in South Africa
    d. Hurricanes on the Horn of Africa
    e. Water pollution in North Africa

4. Deforestation in Sub-Saharan Africa causes which of the following problems?
    a. Moisture loss
    b. Flooding in South Africa
    c. Soil erosion
    d. A and B above
    e. A, B, and C above

5. What is the longest river in Africa and the world?
    a. Congo   b. Niger   c. Nile   d. Senegal   e. Zambezi

6. What is Sub-Saharan Africa's (and the world's) largest desert?
    a. Congo   b. Angolan   c. Kalahari   d. Namib   e. Sahara

7. Why is Sub-Saharan Africa experiencing population growth?
    a. High birth rate coupled with low death rate
    b. High rate of immigration
    c. High rate of urbanization
    d. Preference for large families
    e. All of the above

8. How do the majority of the people of Sub-Saharan Africa earn a living?
    a. Mining
    b. Manufacturing

c. Forestry
d. Fishing
e. Agriculture

9.   In which part of Africa has the impact of HIV/AIDS been most devastating?
      a. North   b. South   c. The Center   d. The Horn   e. The West

10.   What is Apartheid?
      a. A political subdivision in Tanzania
      b. A political party in Ethiopia
      c. A policy of racial separation in South Africa
      d. A landform type found on the west coast of Africa
      e. The Swahili name for bird flu

11.   More than anything else, which of the following has linked Africa to the Americas and
      Europe?
      a. Cocoa
      b. Foreign aid
      c. Language
      d. Rubber
      e. Slavery

12.   What African language became the most widely spoken Sub-Saharan language in the region?
      a. Mandingo
      b. Yoruba
      c. Amharic
      d. Swahili
      e. Igbo

13.   What is the significance of Nubia
      a. It was the only country of Sub-Saharan Africa to escape colonization
      b. It was the first significant state to emerge in Sub-Saharan Africa
      c. It was the most populous country of Africa
      d. It is the leader of the African Union
      e. It is the largest country in Sub-Saharan Africa

14.   What happened at the Berlin Conference?
      a. The African Union was founded
      b. African countries agreed to a guest-worker agreement with the European Union
      c. European colonial powers divided Africa among themselves
      d. African and European countries drafted a peace treaty to end the conflict between the
          two regions
      e. South Africa decided to end Apartheid

15. Which of the following have been identified as reasons for poverty in Sub-Saharan Africa?
    a. Unfavorable environmental conditions
    b. Slavery
    c. Failed development policies
    d. Corruption
    e. All of the above

Answers: 1-A; 2-B; 3-A; 4-E; 5-C; 6-E; 7-D; 8-E; 9-B; 10-C; 11-E; 12-D; 13-B; 14-C; 15-E

Chapter 7
SOUTHWEST ASIA and NORTH AFRICA

LEARNING OBJECTIVES
- This chapter introduces Southwest Asia and North Africa, a region commonly known as the Middle East.
- While this region is usually associated with arid climates, oil, and Islam, the student should be aware that the region is far more complex than this.
- The student should understand the centrality of this region to world trade and religion.
- This chapter provides insights into many ongoing political conflicts in the region.
- At the conclusion of this chapter, the student should be familiar with the physical, demographic, cultural, political, and economic characteristics of Southwest Asia and North Africa.
- In addition, the student should understand the following concepts and models:

  · Islamic fundamentalism
  · OPEC (Organization of Petroleum Exporting Countries)
  · Maghreb and Levant
  · Salinization
  · Pastoral nomadism
  · Monotheism
  · Balfour Declaration

CHAPTER OUTLINE
I. Introduction

This region includes 22 countries, and over 400 million people, and is located on the historic meeting ground between Europe, Asia, and Africa. The region contains some extremely rugged and varied terrain, including deserts and river valleys, and boasts several important *culture hearths* (areas of historical cultural innovation). Innovations that arose in this region have diffused far beyond its boundaries. This region has deep religious significance for Jews, Christians, and Muslims and has also spawned some deep-seated conflicts as a result. The rise and diffusion of *Islamic fundamentalism* (also called revivalism, this movement advocates a return to more traditional practices within the religion and challenges the encroachment of global popular culture) has been an important development in world politics. The presence of vast oil reserves has also placed Southwest Asia and North Africa in a key position with respect to the world's economy*; OPEC:* Organization of Petroleum Exporting Countries

II. Environmental Geography: Life in a Fragile World
A. General Description (Fig. 7.4): this region has great variation, including rocky plateaus, mountain ranges and desert, and climates from arid to well-watered
B. Legacies of a Vulnerable Landscape (Fig. 7.5)
1. One of the world's early culture hearths, the land in this area has

been intensively used, causing many problems and destruction

2. Deforestation and overgrazing: forest rimming the Mediterranean mostly gone; famous cedars of Lebanon now just a few scattered groves; causes: overgrazing, other intensive uses; forests are slow-growing, vulnerable to fire, don't replace themselves easily

3. *Salinization:* buildup of toxic salts in the soil, a side effect of irrigation, which has been used in this region for centuries; fresh water contains small amounts of dissolved salts; irrigation puts this water and salt on farmlands; lack of rainfall means there's no way to flush out the salts; gradually salts become concentrated, ruining fields; especially a problem in Iraq, central Iran

4. Managing water includes modifying drainage systems and water flows; *Qanat system:* Iranian practice of tapping into groundwater through a series of gently sloping tunnels; other adaptations include waterwheels, flood irrigation, canal systems; Aswan High Dam in Egypt (built 1970) increased water storage and generates electricity, but also increases salinization, sedimentation of Lake Nasser, and cases of schistosomiasis (parasitic disease carried in water)

5. *Fossil water:* water supplies stored underground during earlier wetter periods tapped in Libya, moved 600 miles to north coast to expand farmlands

6. *Hydropolitics:* the interplay of water resource issues and politics; it has raised tensions between countries that share drainage basins (see Fig. 7.12)

C. Regional Landforms
1. *Maghreb:* translation from Arabic is "western island" (Morocco, Algeria, Tunisia); Atlas Mountains on Mediterranean east coast are related to Alps in Europe; topography varies from mountains to sand

2. *Levant:* Eastern Mediterranean region, including Lebanon; mountains within 20 miles of the Ocean; Arabian Peninsula is a plateau, sloping eastward; rugged highlands in Oman and Yemen; drifting dune fields elsewhere

3. Iranian and Anatolian plateaus (Iran, Turkey); *Anatolia:* the large peninsula of Turkey, also sometimes called Asia Minor; average height 3,000–5,000 feet, prone to earthquakes; Elburz Mountains in northern Iran are >18,000 feet

4. *Mesopotamia:* land between the Tigris and Euphrates rivers in modern-day Iraq

5. Jordan River Valley straddles borderlands of Israel, Jordan, Syria, and drains into the Dead Sea

D. Patterns of Climate (Fig. 7.11)
1. Complicated climate because of latitude and altitude

2. Driest conditions in North Africa in the Sahara; central Sahara gets less than 1 inch rain per year

3. Mediterranean climate in Morocco, Algeria, Tunisia (hot, dry summers, cooler, moist winters); also along the coastline

of the Levant, Syria, Turkey, and northwest Iran

E. The Uncertainties of Regional Climate Change
  1. Climate change will aggravate existing environmental problems, especially water shortages
  2. Rising sea level could cause loss of farmland in the Nile Delta

III. Population and Settlement: Changing Rural and Urban Worlds

A. Population density within the region varies with availability of water

B. The Geography of Population (Fig. 7.14)
  1. More than 400 million people in this region
  2. *Physiological density* (the number of people relative to the amount of farmable land) in this region is among the world's highest; most people live in moister areas, with two dominant population clusters in the Atlas Mountain region, and Egypt's Nile River Valley

C. Water and Life: Rural Settlement Patterns (Fig. 7.14)
  1. This region is one of the world's earliest hearths of *domestication* (which occurs when plants and animals are purposefully selected and bred for their desirable characteristics); wheat, barley, cattle, sheep, and goats are important; much of the earliest agricultural activity was located in the *Fertile Crescent* (an ecologically diverse zone stretching from the Levant through Syria and into Iraq)
  2. *Pastoral nomadism:* a traditional form of subsistence agriculture in which practitioners depend on the regular and systematic seasonal movement of livestock for a large part of their livelihood; some nomads practice *transhumance* (seasonal movement of livestock to cooler, greener high country pastures in the summer, and returning to valleys and lowlands in the winter); fewer than 10 million pastoral nomads are in the region today
  3. Oasis life: oases may be either natural (high groundwater levels) or human-made (relying on deep-water wells); oases dominated by close-knit families who work their own subsistence plots or work for absentee landowners; most oases are small and include trade, subsistence, and commercial crops
  4. Settlement along *exotic rivers* (rivers that come from distant, moister regions, in this arid zone, bringing precious water and nutrients from those distant lands); examples: Nile, Tigris, Euphrates; exotic rivers make settlement possible, but they're vulnerable to overuse; higher population densities possible with exotic rivers than under nomadism
  5. *Kibbutzs:* collectively worked settlements in Israel that produce grain, vegetable, and orchard crops; these are made possible by exotic rivers
  6. The challenge of dryland agriculture: this region has a mix of varied crops and livestock, drought-resistant trees (olives, almonds, citrus); vineyards, barley, and wheat where possible; irrigation and mechanization widespread

D. Many-Layered Landscapes: The Urban Imprint
   1. A long urban legacy: city life began in Mesopotamia (Iraq) by 3500 B.C., in Egypt by 3000 B.C.; early cities were centers of politics and religion; trade centers at the junctions of caravan routes and on seaports began to arise around 2000 B.C.
   2. Islam influenced city form, with the most common form being the *medina* or walled urban core; European colonists also left their imprint on architecture
E. Signatures of Globalization
   1. Urban centers are focal points of economic growth
   2. Rural people drawn to cities
   3. Growth in several cities has been rapid in recent years: Algiers (Algeria) and Istanbul (Turkey) populations doubled in recent years; squatter settlements are a problem; wealthier people moving to outlying areas (urban flight); Cairo (Egypt) has more than 15 million people
   4. Oil-rich states more urban than others (80 percent), have modern architecture, extensive transportation infrastructure, mosques, modern office buildings, futuristic structures, and oil refineries
F. Recent Migration Patterns: global economy and recent political events have sparked migration; (1) rural-to-urban shift; (2) migration within the region for job opportunities (today more than 85% percent of the Saudi workforce are foreigners who export more than $20 billion in wages as remittances to their homelands; (3) residents from the region migrate elsewhere for jobs, with Turkish guest workers moving to Germany, Algerians and Moroccans mostly to France, others to North America; (4) wealthy residents left politically turbulent areas like Lebanon and Iran; (5) Jewish immigration to Israel is rising, especially after breakup of the former U.S.S.R.
G. Shifting Demographic Patterns
   1. Population growth is still a key issue; there is greater variation in growth rates throughout the region
   2. Smaller families and lower total fertility rates are becoming more common in many countries of this region (Tunisia, Iran, Turkey)
   3. Some places, like the West Bank, Gaza (in Israel), Iraq, and Yemen have the highest TFRs in the region
   4. Continued environmental challenges persist, with limited resources and a large proportion of citizens younger than age 15
   5. Birth rates are likely to decline in the future

IV. Cultural Coherence and Diversity: Signatures of Complexity
A. Patterns of Religion
   1. Hearth of the Judeo-Christian tradition: Judaism and Christianity trace their roots to the eastern Mediterranean part of this region; *monotheism:* a belief in one God is an important part of Judaism, Christianity, and Islam; Christianity is an outgrowth of Judaism, based on teachings of Jesus and his followers; Coptic Christians in Egypt; Maronites in the Levant
   2. The emergence of Islam (Fig. 7.24): Islam originated in Southwest Asia in A.D. 622; it is

a continuation of the Judeo-Christian tradition; Islam holds that both the Hebrew Bible (Old Testament) and the Christian New Testament are basically accurate (though incomplete), and considers Abraham, Moses, and Jesus to be true prophets; *the Quran (Koran),* a book of revelations received by Muhammed from Allah (God), is Islam's holy book; *Islam* means "submission to the will of God"; five pillars of Islam: (1) repeat basic creed ("There is no God but God, and Muhammed is his prophet"); (2) prayer facing Makkah (Mecca) five times daily; (3) giving charitable contributions; (4) fasting between sun-up and sun-down during the month of Ramadan; (5) making at least one religious pilgrimage *(Hajj)* to Makkah (Muhammed's birthplace); split in Islam over succession after Muhammed died: Shiites favored passing power to Muhammed's son-in-law, Ali (who was later martyred); Sunnis wanted to pass power through established clergy; today Iran is a theocratic state (religious leaders guide policy)

3. Diffusion of Islam (Fig. 7.22): quick diffusion, along trade and military routes; Persia, eastern Roman Empire, North Africa, Spain, and Portugal were Muslim by 750 A.D.; in the 1400s, the Ottoman Empire (centered on modern-day Turkey) adopted Islam, while Iberia (Spain and Portugal) became Christian

4. Modern religious diversity (Fig. 7.24): Sunni Muslims are a majority in the region, except in Israel (Judaism), and Cyprus (Greek Orthodox); Shiite Muslims dominate in Iran, southern Iraq, Lebanon, Sudan, and Bahrain; Israel's population is 80 percent Jewish; the city of Jerusalem is important to three religions (Judaism, Christianity and Islam); Coptic Christians make up 7 percent of Egypt's population

B. Geographies of Language
1. Semites and Berbers: Arabic-speaking Semitic peoples are found from Morocco to Saudi Arabia, spread with the diffusion of Islam; Hebrew, spoken by Jews and in Israel, is also a Semitic language; Berber, an older Afro-Asiatic language, is found in the Atlas Mountains and in the Sahara region
2. Persians and Kurds (Indo-European languages): Persian dominates in the Iranian Plateau region, Farsi is the dominant Persian dialect; Kurdish speakers (Kurds) are found in northern Iraq, northwest Iran, and eastern Turkey (where the three countries come together); Kurds have faced very serious and sometimes deadly discrimination; "Kurdistan" has been called the "world's largest nation without its own political state"
3. The Turkish imprint: Turkish languages are part of the Altaic family (origin in Central Asia); related Altaic languages in other regions include Azeri, Uzbek, Uighur

C. Regional Cultures in Global Context
1. Islamic internationalism: in addition to this region, Muslims

are found in central China, European Russia, central Asia, southern Philippines, Malaysia, and Indonesia (Indonesia has the world's largest Muslim population.); Muslim congregations also growing in urban centers in Europe and North America; Islam is a global religion

2. Globalization and cultural change: access to modern communication technology brings world culture to this region; expansion of Islamic fundamentalism is partly a reaction to threat posed by external cultural influences; technology (including the Internet and cell phones) is becoming increasingly important in the region, especially to the wealthy and/or well-educated; former colonial relationships have evolved into guest-worker situations (Algerians in France, for example); many people from outside the region (Jews and Christians) make pilgrimages to the Holy Land

V. Geopolitical Framework: Never-Ending Tensions (Fig. 7.30)

A. The Colonial Legacy
1. European colonization came after WWI (1918); before this, Ottoman Turks—imperialists within the region —repelled Europeans; European imprint remained after they left in the 1950s

B. Imposing European Power
1. The French came to Algeria in 1830; French government wanted Algeria to become a key part of France; France also established protectorates in Tunisia (1881) and Morocco (1912), ensuring French dominance in the Maghreb; after France defeated Germany/Ottoman Turk alliance in WWI, it gained land in the Levant (Syria and Lebanon)

2. Great Britain gained control of small coastal states to secure sea lanes from India; these states included Kuwait, Bahrain, Qatar, United Arab Emirates, Aden (Yemen); British-engineered Suez Canal in Egypt linked the Mediterranean and the Red seas in 1869, taking more direct control of Egypt and Sudan in 1883; After WWI, Britain engineered creation of Saudi Arabia, through the Saud family, and placed Saud family in control; Britain also created Iraq from three dissimilar Ottoman provinces: (1) Basra (Arabic-speaking, Shiites); (2) Baghdad (Arabic-speaking Sunnis); (3) Mosul (Kurdish-dominated)

3. Other European colonies: Italians in Libya, Spanish in Morocco

4. Persia (Iran) and Turkey were never directly colonized; Britain and Russia had spheres of economic influence in Persia, which was technically independent; Turks, led by Mustafa Kemal (Ataturk or Father of Turks), assured Turkey's independence against French and Greeks in 1920s and set Turkey on a modernist (Western-looking) course when he became Turkey's leader

5. Decolonization and independence: Europeans began to leave the region before WWII; by the 1950s, most countries in the

region were independent; Algerians fought for independence from France from 1954–1962; Iraq gained freedom from Britain in 1932, but conflicts resulting from British-drawn borders persist; French partitioning of Lebanon and Syria caused later internal conflicts

C. Modern Geopolitical Issues

1. Across North Africa: Morocco has invaded and annexed western Sahara; Muammar al-Qaddafi of Libya encouraged tensions for years, but recently has vowed to disarm and campaigns for African Unity, earning praise from the United States and E.U.; Islamist conflicts in Algeria; Morocco and Egypt experience instability created by Islamist elements; Sudan faces conflicts, most recently in Darfur.

2. The Arab-Israeli conflict (Fig. 7.32): The Balfour Declaration created the state of Israel as a homeland in Palestine for Jewish people in 1948; conflict arose immediately after British left as Arab Palestinians rejected this; Israel's land area has grown over the years; many Palestinians have fled to neighboring countries; there have been three wars in the region (1956, 1967, 1973); Jewish immigrants from around the world (including Russia, Europe, and the United States) have settled in Israel, increasing tensions, not only between Palestinians and Israelis, but also between Israel and other countries (e.g., Syria, Lebanon); new initiative to give back some land to Palestinians in 2005

3. Devastated Iraq: U.S. invasion in 2003; handover of power to Iraqis on June 28, 2004; elections in February 2005; much uncertainty remains; Iraq's different regions have different cultures; U.S. troops and contractors remain.

4. Instability in Saudi Arabia: Saudi Arabia is not a democracy; a growing number of Saudi citizens are pressing for greater freedom; a bombing in 2003 shook the country and has affected oil prices; the future is uncertain; 15 of 19 September 11 hijackers were Saudi citizens.

5. The non-Arab fringe: Turkey and Iran. Turkey is committed to join the EU; freedoms are increasing, but pressures remain. Iran oscillates between reform and fundamentalism; United States, EU are concerned about Iran's nuclear program

D. An Uncertain Political Future

1. Complex relationship between United States and countries of this region: Israel and Turkey are U.S. allies; Iran, Iraq, Syria, Libya sometimes seen as enemies; most countries in the region have many U.S.-made weapons.

2. The outcome of the current war in Iraq and the situation in Saudi Arabia are still uncertain

3. Democracies in the region are scarce (Turkey and Israel)

4. Oil in the region—and its continuing importance to world economic development —mean that this region will continue to be important, especially as rising demand for oil causes prices to rise.

VI. Economic and Social Development: Lands of Wealth and Poverty

A. The geography of fossil fuels (Fig. 7.36)

1. Oil is unevenly distributed in the region: Saudi Arabia, Iran, United Arab Emirates, Libya, Algeria, Bahrain, Qatar, Kuwait all produce oil for export; Morocco, Sudan, Israel, Jordan, Lebanon have little oil; Turkey produces oil, but must import more to meet its own needs

2. This region has 69 percent of the world's proven oil supplies (only 7 percent of its population); Saudi Arabia has a population of only 28 million, but has 21 percent of the world's proven oil reserves

B. Regional Economic Patterns

1. Fluctuations in oil prices cause shifts in economic well-being

2. Higher-income oil exporters (Saudi Arabia, Kuwait, Qatar, Bahrain, United Arab Emirates) have wealth from oil, but not everyone benefits: rural Shiite Muslims in eastern Saudi Arabia less well off, while foreign workers (from Jordan, Egypt, Yemen, Pakistan, Bangladesh, and Philippines) earn low wages; some countries diversify their economies, but many do not

3. Lower-income oil exporters (Algeria, Libya, Iraq, Iran) have stifled economic growth: many Algerians migrate to Europe (France) for better economic opportunities; Libya's political hostility toward other countries, its reputation as an "outlaw" nation, and its heavy military buildup limit economic prosperity; sanctions against Iraq eliminate most oil exports and living standards are extremely low; Iran's leaders limit international trade, limiting Iran's economic prosperity

4. Prospering without oil: Israel has the highest standard of living in the region, with agriculture and industry, including computers and telecommunications products, but poverty is widespread in Palestine; Turkey has a diversified economy with agriculture (cotton, tobacco, wheat, fruit) and industry (textiles, food, chemicals); Istanbul is a regional finance and investment center and the most important tourist destination in the region, and Turkey is also trying to become a member of the European Union; Tunisia and Cyprus have agriculture and tourism; Lebanon used to be the most prosperous state in the region, and has a growing economy with tourism and telecommunications

5. Regional patterns of poverty: civil war has ruined Sudan's economy, destroying infrastructure and creating food shortages in spite of fertile soils; in parts of Morocco, poverty and illiteracy are widespread—this country and others suffer from *brain drain,* as many of the best and brightest seek their fortunes outside Morocco; Egypt's prospects are unclear, with some improvements, but much poverty remaining; it's the region's most populous state at 70 million; Yemen is the poorest in the region, largely rural and

practicing pastoral nomadism and subsistence agriculture

C. Issues of Social Development
1. Varied regional patterns: Israel has high standard of living, but members of Israel's Jewish majority are appreciably better off than the sizable Muslim minority in Israel; even though Saudi Arabia is a wealthy country, one-quarter of its people are illiterate and infant mortality four–five times higher than in Israel
2. A woman's changing world: this region has the world's lowest female labor participation; Turkey's women most free among Islamic countries in the region, but they still rarely work in retail; Saudi women prohibited from driving cars; Iranian women wear the hijab (veil); educational opportunities for women are improving, and it is generally available, but may be segregated by sex; in Libya, more women than men graduate from college; 70% of Algeria's lawyers and 60% of its judges are women, and women are well-represented in the medical profession

D. Global Economic Relationships
1. OPEC's changing fortunes: OPEC can no longer control prices globally, it still influences the price and availability of oil; petroleum and petrochemicals bring in 90 percent of Saudi Arabia's export income; foreign investment of Gulf Arabs tops $800 billion; non-OPEC countries are less dependent on oil: Turkey exports textiles, food, manufactured goods, and Israel exports diamonds, electronics, machinery parts; Tunisia also exports; the EU is the major trade partner of all three countries.
2. Regional and International Linkages: this region is becoming more closely linked with the European Union (EU), with "Euro-Med" agreements between the EU and Morocco, Tunisia, Jordan, Israel; Turkey is trying to join the EU; Arab countries are not following this lead; the Arab Free-Trade Area (AFTA) was established in 1998, will try to eliminate all intra-regional trade barriers by 2008
3. The Geography of Tourism: Tourism was traditionally based on religious pilgrimages (to Mecca and the Holy Land) and historical explorations, but now increasingly touts beaches and other typical tourist amenities

VII. Conclusions

Southwest Asia and North Africa are located at the crossroads of three continents. This, along with petroleum resources, assures a global role for this region. The colonial experience of this region was not long, but consequences linger in the form of boundary issues. This region has significance to three of the world's religions: Judaism, Christianity, and Islam, which share common philosophical as well as geographical roots in this region. In the 80s and 90s, the rise and diffusion of Islamic fundamentalism has been an important influence in the region. Much of the region's economy is based on petroleum and related products. For this reason, the economy is prone to follow the fortunes of the oil industry.

PRACTICE MULTIPLE CHOICE QUIZ

1. Why is the region of Southwest Asia and North Africa considered to be a world culture hearth?
    a. It was a seedbed of agricultural domestication
    b. Judaism was founded in the region
    c. Islam was founded in this region
    d. Christianity was founded in this region
    e. All of the above

2. What is the major reason for environmental problems in Southwest Asia and North Africa?
    a. Domination of heavy industry in the region
    b. High population density
    c. The petroleum industry
    d. The long history of human settlement in the region
    e. All of the above

3. What causes salinization (the buildup of toxic salts in the soil)?
    a. Industrial pollution
    b. Petroleum refining
    c. Petroleum extraction
    d. Salt mining
    e. Extensive use of irrigation in desert lands

4. All of the following statements about climate in Southwest Asia and North Africa are true, EXCEPT…
    a. Aridity dominates large portions of the region
    b. Plants and animals have adapted to extreme conditions in the region
    c. Some of the driest parts of the region are found in North Africa
    d. The Arabian peninsula is moister than the rest of the region
    e. The region is often called, *the dry world*

5. What is the physiological density of Southwest Asia and North Africa, and why is this important?
    a. The physiological density is high, which means that the number of people per square mile is high
    b. The physiological density is high, which means that the number of people per unit of arable land is among the highest on Earth
    c. The physiological density is low, meaning that the number of livestock is relatively low compared to the number of people
    d. The physiological density is low, meaning that there is ample farmland to support the population
    e. The physiological density is moderate, meaning that there is a modes number of people per square kilometer

6. Which of the following are among the most common livelihoods in Southwest Asia and North Africa?
   a. Irrigated agriculture along exotic rivers
   b. Oasis agriculture
   c. Pastoral nomadism
   d. A and B above
   e. A, B, and C above

7. All of the following statements about cities in Southwest Asia and North Africa are true, EXCEPT...
   a. European colonialism added another layer to the urban landscape features
   b. Many cities reflect Islamic influences
   c. Modern popular culture has significantly altered traditional architecture throughout the region
   d. Cities in the region have been the focal points for local and long-distance trade
   e. Cities in the region served as centers of political and religious activity

8. What pattern of migration is found in Southwest Asia and North Africa?
   a. Large numbers of workers are migrating within the region to areas with job opportunities
   b. Political forces and war have encouraged migration
   c. Residents of the region are migrating to other parts of the world, especially to Europe
   d. Rural-to-urban migration is occurring
   e. All of the above

9. What kind of government does Iran have?
   a. Theocracy
   b. Representative democracy
   c. Democratic Republic
   d. Constitutional monarchy
   e. Communist dictatorship

10. Followers of which religion form the majority population in al of the countries of Southwest Asia and North Africa except Israel?
    a. Buddhism
    b. Coptic Christianity
    c. Islam
    d. Judaism
    e. Orthodox Christianity

11. What was the major reason for the late arrival of European colonization of Southwest Asia and North Africa?
    a. The domination of the region by the Ottoman Empire (Turks)
    b. Military resistance to European countries by Persia
    c. The difficult physical geography of the region
    d. The lack of valuable resources in the region
    e. The presence of serious illnesses in the region and Europeans' lack of resistance (as in Africa)

12. What country of Southwest Asia and North Africa is the only one that does not have a majority Muslim population?
    a. Egypt   b. Israel   c. Libya   d. Morocco   e. Turkey

13. Which country of Southwest Asia and North Africa functioned as a democracy for many years?
    a. Israel
    b. Saudi Arabia
    c. Turkey
    d. A and B above
    e. A and C above

14. To what do the most prosperous countries of Southwest Asia and North Africa owe their wealth?
    a. Massive oil reserves
    b. Information technology
    c. Industry
    d. Banking services
    e. Export manufacturing

15. What is the only country in Southwest Asia and North Africa to seek membership in the European Union?
    a. Iraq   b. Israel   c. Libya   d. Turkey   e. United Arab Emirates

Answers: 1-A; 2-D; 3-E; 4-D; 5-B; 6-E; 7-C; 8-E; 9-A; 10-C; 11-A; 12-B; 13-E; 14-A; 15-D

Chapter 8
EUROPE

LEARNING OBJECTIVES:
- This chapter introduces Europe, one of the world's most densely settled modern regions, with a population of more than half a billion people
- An important objective of this chapter is to understand the dynamics of nationalism that threw the region into armed conflict twice during the 1900s, and many times before that.
- In addition, this chapter sheds light on the changing dynamics that have resulted in the recent formation of the European Union and the evolution of a common currency, the euro.
- At the conclusion of this chapter, the student should be familiar with the physical, demographic, cultural, political, and economic characteristics of North America.
- In addition, the student should understand the following concepts and models:

  · Balkanization
  · Cold War
  · Command economy
  · European Union
  · Feudalism
  · Privatization
  · Secularization

CHAPTER OUTLINE
I. Introduction
Europe is one of the most diverse places on Earth. In an area smaller than North America, Europe has a variety of climates, landforms, and agricultural output. This region has over 500 million people and 40 countries, with many different languages and cultures. Because of these differences, Europe has a long history of warfare, mostly occurring along nationalist lines. Europe is now globalizing, with the formation of a supranational political and economic entity, the *European Union (EU)*. Still, national identity has led to the dissolution of countries in eastern Europe, most notably Yugoslavia. Europe has had a large impact on the world. It is the cradle

of Western civilization and the Industrial Revolution. European countries colonized countries in every region of the world.

II. Environmental Geography: Human Transformation of a Diverse Landscape
A. Introduction
   1. Europe's environmental diversity is great, despite its small land area
   2. Four factors explain this diversity: (1) complex geology with both newest and oldest landscape types; (2) Europe's latitudinal extent (distance from equator), from arctic to Mediterranean subtropics; (3) modification of latitudinal controls by interaction of land masses and large bodies of water

(Atlantic Ocean, Black, Baltic, and Mediterranean seas); (4) thousands of years of humans living in Europe has fundamentally changed its landscape

B. Landform and Landscape Regions (Fig. 8.1)

1. European lowland (a.k.a. North European Plain): southwest France to Poland, plus southeast England; focus of western Europe, with high population density, agriculture, major cities, industries, major rivers

2. Alpine mountain system: "spine of Europe," east-west series of mountains from the Atlantic Ocean to Black Sea; formed 20 million years ago; (1) *Pyrenees* form border between Spain and France, peaks top 10,000 feet; (2) Alps run 500 miles from France to Austria; highest in West (more than 15,000 feet at Mt. Blanc); (3) Appenines, the "spine of Italy," south of the Alps; (4) Carpathians at eastern edge of Alpine system, under 9000 feet (eastern Austria to Romania and Yugoslavia)

3. Central Uplands: between Alps and Lowland, from France to Czech Republic; have iron, coal, and other raw materials for manufacturing

4. Western Highlands (from Portugal to Finland, plus part of British Isles): *fjords* (flooded U-shaped valley coastlines) found in north; this *shield landscape* contains 600 million year old rocks (world's oldest) exposed by recent glaciation

C. Europe's Climate (Fig. 8.7)

1. Climates here moderated by the *North Atlantic Current,* a warm water current from North America's Gulf Stream; Europe has three climate types

2. Marine west coast climate along the Atlantic Coast: no winter months below freezing, but cold rain, snow common; summers often cloudy with drizzle or rain (example: Ireland)

3. Continental climate inland: hotter summers, colder winters, one–two months average below freezing; rainfall sufficient for non-irrigated farming

4. Mediterranean climate in southern Europe (Spain to Greece); dry summer season that may include drought; irrigation common

D. Europe's Ring of Seas

1. EuroBaltic, English Channel, Mediterranean, Black, North (which has good fishing and abundant oil and natural gas)

2. Rivers and ports: Many rivers are navigable and connected by canals to handle barge traffic throughout Europe; rivers include Seine, Rhine, Elbe, and Danube (Europe's longest); Europe's port cities: Rotterdam (Rhine); Le Havre (Seine); Hamburg (Elbe); London (Thames); Gdansk (Wisla)

E. Environmental Issues, Local and Global, East and West (Fig. 8.4)

1. Europe's long history of agriculture, resource extraction, industrial manufacturing, and urbanization created serious environmental problems (air and water pollution, acid rain)

2. European Union has addressed environmental issues since 1970s with successful regional solutions; western Europe has one of world's "greenest" (best) environments
3. European voters support environmental legislation in Europe and globally, including reduction of greenhouse gases
4. Eastern Europe, under Soviet-style communist economic system (1945–90), neglected its environment; Soviet-designed nuclear power plants in eastern Europe may be prone to accidents
5. Solving eastern Europe's environmental problems is complicated by challenges of its ongoing economic and political evolution

F. Global Warming in Europe: Problems and Prospects
1. Dwindling sea ice, sparse snow cover is leading to droughts in the Mediterranean
2. Longer growing seasons may help northern areas
3. Europe and Kyoto Protocol: collected goal of 8% below 1990 GHG levels, with most developed countries having the strictest goals
4. EU's Emission Trading Scheme: Each plant may meet its GHG emission goal or buy credits from plants that are under their GHG allowance; plan will be revised
5. EU only 10% toward its goal

III. Settlement and Population: Slow Growth and Rapid Migration (Fig. 8.14)

A. Population Density in Core and Periphery

1. Half a billion people live in Europe
2. Highest densities in historical industrial core (England, Netherlands, northern France, northern Italy, western Germany)

B. Natural Growth: Beyond the Demographic Transition
1. Europe is at the last stage of the demographic transition; birthrates lower than death rates; immigration prevents population loss; some European countries offer financial incentives to couples who have children;
2. Causes of decline: women working, widespread contraception, lack of affordable housing
3. Low birthrate countries are giving incentives to encourage more births

C. Migration to and within Europe (Fig. 8.15)
1. Resistance to unlimited immigration into Europe growing: scarce jobs should go to Europeans first; concerns about foreign terrorism and dilution of national culture
2. Immigrants from former European colonies in Asia, Africa, the Caribbean; also from former Soviet Union, Eastern Europe; France and Germany's populations are 10% immigrants.
3. European Union working toward common immigration policy by year 2004
4. *Guest workers:* immigrants brought into a country to do work, usually low-wage; Germany has Turkish guest

workers; other European countries have migrants from their former colonies in Africa, Asia, Caribbean

D. The Geography of "Fortress Europe"
   1. *Schengen Agreement:* agreement between members of the European Union that allows people to travel within the member states without needing a passport; it's brought about strict border controls between EU and non-EU countries, forming "Fortress Europe"
   2. Difficulties with the Schengen Agreement: Spain and Italy have long coastlines and carry the burden of policing the outer boundaries of the Shengen Agreement's territory *("Shengenland")*; illegal immigrants from Asia and Africa are dropped on beaches in Italy and Spain and then travel to northern Europe

E. The Landscapes of Urban Europe
   1. Europe is highly urbanized—more than 50 percent in most countries; 90 percent in United Kingdom and Belgium. Historical momentum: spread of cities in Europe associated with classical empires (Greece, Rome), which used rivers as settlement points because of their transportation potential
   2. The past in the present: medieval (900–1500 A.D.) cities densely settled; buildings right next to streets, green space only near churches and public squares; Renaissance-Baroque (1500–1800): wider streets, large gardens, monuments, more open spaces; industrial (1800–present): walls and fortifications removed, factories and industrial areas built on edge of the city, and urban sprawl developed

F. Protecting the Sense of Place
   1. Europeans take strong sense of cultural identity from their cities, and protect urban landscapes
   2. Europeans protect skylines by prohibiting building of high-rises in central cities, pushing them to outlying areas. Preservation of historic buildings is an important part of the European experience and includes the protection of skylines and historical preservation

IV. **Cultural Coherence and Diversity: A Mosaic of Differences**
   A. Differences in languages, customs, religions have created local and regional identities that caused conflict
      1. European cultures have influenced many world regions through colonialism and imperialism (globalization)
      2. Many European cultures resist U.S.-influenced popular culture (France, for example)
   B. Geographies of Language (Fig. 8.22)
      1. Most Europeans learn multiple languages
      2. Germanic languages: 200 million northern Europeans speak Germanic languages (German, English, Dutch, Norwegian, Swedish, Danish, and Icelandic)
      3. Romance languages: 200 million southern Europeans speak dialects of Latin (Italian, French, Spanish, Portuguese, Catalan, Romanian)

4. Slavic languages: 80 million eastern Europeans speak Slavic languages (Polish, Czech, Slovakian, Serbo-Croatian, Bulgarian, Slovenian); countries with Roman Catholic ties use the Roman alphabet (as we do in the United States); those with ties to Eastern Orthodox Church use Cyrillic alphabet
5. Minor Indo-European languages: Celtic languages (Irish, Scots, Welsh, Breton); Hellenic languages (Greek); Baltic languages (Latvian, Lithuanian);
6. Non Indo-European languages: Uralic (Hungarian); Finnish and Sami; Altaic (Turkish, spoken in parts of Bulgaria); Basque (small part of Spain – there is a Basque separatist movement in Spain)

C. Geographies of Religion, Past and Present (Fig. 8.24)
1. The schism between Western and Eastern Christianity: when Greek missionaries refused to accept the Roman Catholic hierarchy and control by Roman bishops, the two groups split in A.D. 1054; Roman Catholics use Latin alphabet; Eastern Orthodox churches used Cyrillic (Greek) alphabet
2. The Protestant Revolt: began in 16th century; has divided the region since
3. Conflicts with Islam: in the east, Ottoman Turks expanded Islam into the Balkans; on the west, Moorish (Moroccan) influence in 12th century; Muslims fairly tolerant of religion in lands they conquered, but Christians not tolerant of Islam
4. A geography of Judaism: Jews in Europe were forced from Palestine during the Roman era, settled in Moorish Spain; when Christians conquered this Muslim-controlled area, they expelled Jews; many settled in an area of eastern Europe called "the Pale;" Nazis from Germany persecuted Jewish people in their concentrated settlements and killed around 6 million, while others suffered in concentration camps
5. Patterns of contemporary religion: Europe becoming more *secularized*; Protestantism in northern Germany, Scandinavia, England; Catholicism in Spain, Ireland, France, Italy, Austria, southern Germany; conflict in Northern Ireland between Irish nationalists (mostly Catholic) and British loyalists (mostly Protestant) has religious aspects

D. European Culture in Global Context
1. Globalization and cultural nationalism: since WWII, Europe has been inundated with American culture (music, television, consumer goods); U.K., Italy, Hungary accept it; France and Germany try to prevent it by subsidizing indigenous films and (in France) creating "academies" to keep English terms from entering the language
2. Migrants and culture: influx of Muslim migrants (4.5 million in France; 2.5 million Muslim

Turks in Germany) is adding a new cultural element; ethnic clustering and *ghettoization* growing common; far right-wing nationalist parties on the rise (e.g., skinheads, neo-Nazis)

V. Geopolitical Framework: A Dynamic Map (Fig. 8.29)

A. There are 40 independent states in the small area of Europe; Europe invented the *nation-state:* completely new geopolitical and spatial entity fostered by ethnic, cultural nationalism; though this region has seen two major wars since 1900, its people now seek unity

B. Redrawing the Map of Europe through War (Fig. 8.30)
1. In the 20th century, two world wars changed Europe's borders
2. World War I (1914–1918): two groups of opponents: France, Great Britain, Russia vs. Germany, Italy, and Austria-Hungary; Treaty of Versailles ended the war, punished the losers, and created different countries for different nations of people; encouraged *irredentism:* state policies designed to reclaim lost territory (real or imagined)
3. In 1930s Depression era, three competing ideologies emerged: (1) Western democracy and capitalism; (2) Soviet-style communism; (3) fascist totalitarianism in Germany and Italy
4. World War II era (1939–1945): Germany began to annex much of continental Europe; again, two groups of opponents:

Germany and Italy (Axis) vs. Britain, France, U.S.S.R., United States (Allies); Axis surrendered in 1945; Allies divided Europe and the Cold War began

C. A Divided Europe, East and West
1. *Cold War* geography: after World War II, Russia (U.S.S.R.) occupied eastern Europe, Western Allies occupied western Europe, and the city of Berlin and Germany were divided; U.S.S.R. occupied adjacent countries to make a *buffer zone,* a region that would protect Russia from further European invasion; border between east and west was closed off by what came to be called the *Iron Curtain* (with the exception of the concrete Berlin Wall, this was not a physical entity, but rather a political idea)
2. *Cold War* meant a constant threat of war between two alliances, North Atlantic Treaty Organization (NATO) in western Europe, and Warsaw Pact countries of eastern Europe; both sides stockpiled weapons and waged a propaganda war, raising fear between the groups
3. The Cold War thaw: began in 1989, when Poland elected a non-Communist leader; political instability in the U.S.S.R. and a desire for economic and political restructuring in the countries of eastern Europe were key causes of the thaw; revolutions in Warsaw Pact countries were nonviolent, except in Romania;

one outcome was a revival of national feelings in the region, with Czechoslovakia dividing peacefully, while Yugoslavia divided as a result of armed conflicts; Germany reunited.

4. A New Geopolitical Stability? eastern European countries have been accepted into Europe as a whole; many former Warsaw Pact nations are now NATO members, and NATO has a warmer relationship with Russia, especially after the September 11 terrorist attacks

D. The Balkans: Europe's Geopolitical Nightmare (Fig. 8.33)

1. Four different and sometimes contradictory scales of conflict in Yugoslavia: (1) local and regional tensions within the region (Serbia, Montenegro; Kosovo); (2) tensions between countries of the region and its neighboring states (Macedonia, Albania, Bosnia, Croatia, Slovenia); (3) relations with the rest of Europe, with NATO and with the EU; and (4) global implications of the crisis, including external sympathizers with the various ethnic groups involved and peacekeeping organizations, such as the United Nations

2. Some people believe that the ethnic separatism in this region (the Balkans) is inevitable and will always lead to violence and terrorism; the term *balkanization* describes the fragmented geopolitical processes involved with small-scale independence movements and the phenomenon of mininationalism as it develops along ethnic fault lines

4. Historical Background: at the end of WWI Yugoslavia was created from eight different nations with three different religions, five different languages, and two different alphabets; the Yugoslav province of Serbia controlled Yugoslavia between WWI and WWII; Germany invaded Yugoslavia in WWII; people in the Yugoslavian province of Croatia supported the German invasion because it freed them from rule by the Serbs; Serb loyalists and pan-Yugoslav partisans supported the Allies; after WWII, Yugoslavia united as a communist country independent of the U.S.S.R.; unification ended in 1990, when former provinces seceded; after 1990, Slobodan Milosevich, the Serbian nationalist leader of Yugoslavia, wanted to keep Serbs together; Slovenia seceded in 1990; Croatia in 1991

3. More recent events: Macedonia and Bosnia-Herzegovina in 1992; civil wars broke out throughout the country; finally ended with the Dayton Peace Accords in 1995; ethnic tensions remain, and war crimes trials have occurred; Kosovo declared its independence in February 2008

VI. Economic and Social Development: Integration and Transition (Fig. 8.36)

A. Europe's Industrial Revolution (1730–1850)
Industrial revolution brought two fundamental changes:

1. Machines replaced human labor in many manufacturing processes

2. Inanimate (rather than people or animals) energy sources

such as water, steam, petroleum, powered the new machines

3. Centers of change: Yorkshire and Lancashire, England, were the centers of innovation for industrialization of textile production; these cities had water sources to power waterwheels, weak *guilds* (unions), and raw materials (wool and cotton)

4. Locational factors of early industrial areas; improvements in steam technology made waterwheels obsolete; coal was a cheap fuel source so factories were built near coalfields; iron and steel manufacturing became important; London became an important port and financial center

5. Development of industrial regions in continental Europe: first industrial regions were established around 1820 on the French-Belgian border, near coalfields around the Sambre and Meuse rivers; these regions still important today

B. Rebuilding Postwar Europe: Integration in the West

1. Marshall Plan, a U.S. program that helped rebuild western Europe after WWII; part of European Recovery Program (ERP)

2. ECSC and EEC (European Coal and Steel Community and European Economic Community): ancestors of today's European Union; from its start as a coordinated effort to drop coal and steel tariffs, ECSC grew to create common market for member countries

(France, Belgium, Netherlands, Italy, Germany, and Luxembourg)

3. European Community and Union: in 1965, the EEC created a council, court, parliament, and commission and changed its name to European Community (EC); some EFTA countries (Britain, Denmark, Norway) joined the EC at that time; in 1991, the EC became the European Union (EU), and more members joined; EU has a wider mission, which includes common foreign policies and mutual security agreements, greater economic integration, and a common currency, through the Treaty of Maastricht

4. Euroland, the European Monetary Union: on January 1, 1999, 11 of 15 EU members adopted the *euro,* a common currency for business and trade transactions; on January 1, 2002, citizens started using the *euro* in their everyday lives

C. Economic Integration, Disintegration and Transition in Eastern Europe

1. Historically, eastern Europe has been less well developed economically than western Europe and has been under the control of outsiders (Ottomans, Hapsburgs, Germans, and Soviet Russians)

2. The Soviet Plan: the Soviet Russians (communists) redeveloped eastern Europe after WWII as a *command economy* (centrally planned and controlled economy, generally

associated with socialist or communist countries, in which all goods, services, agricultural, and industrial products strictly regulated)

3. The results: collectivization of *agriculture* did not improve food production and food shortages were common; *industry* focused on heavy industry (e.g., steel), but became dependent on cheap fuel and raw materials from the U.S.S.R.

4. Transition and changes since 1991: after U.S.S.R. disintegrated, eastern European countries went through tremendous change; many introduced *privatization* (the transfer of companies owned by the state to private hands); loss of cheap raw materials and fuel from the U.S.S.R. resulted in a drop of industrial and agriculture output; unemployment and inflation have increased

5. Regional disparities within eastern Europe: successful transitions (Czech Republic, Slovenia, Hungary, Poland) never adopted centralized communism, had good transportation links to the West, and developed strong manufacturing centers with a skilled workforce; unsuccessful transitioners (Macedonia, Moldova, Albania) had adopted centralized communism, had few links with the West and were dependent on U.S.S.R. for raw materials, had agriculture-based economies, and experienced internal political conflict; the other eastern European countries are in the middle, with a combination of traits

VII. Conclusions

As the 21st century gets underway, Europe faces a challenge of dealing with two different areas. While western Europe is one of the wealthiest regions on earth, with a progressive approach to environmental concerns, ideas of nationalism seem to be giving way to a new pan-European identity and a region wide currency. Europe still must deal with the influx of immigrants from other regions and the political tension that this entails, as well as addressing political problems in other regions.

Eastern Europe has different challenges. Political strife, economic stagnation, and environmental degradation are problems that eastern Europe must rectify before it can join the West, which is a goal of all eastern European countries.

PRACTICE MULTIPLE CHOICE QUIZ

1. Which of the following climates is ABSENT from Europe?
    a. Continental midlatitude
    b. Highland
    c. Mild midlatitude
    d. Mediterranean
    e. Tropical

2. All of the following statements about bodies of water in and around Europe are true, EXCEPT…
    a. Europe has few navigable rivers
    b. Five major seas circle Europe
    c. Major ports are found at the mouth of most rivers of Western Europe
    d. The English Channel separates the British Isles from continental Europe
    e. The seas surrounding Europe are connected to each other through narrow straits

3. Which of the following is responsible for the environmental problems facing Europe today?
    a. Industrial manufacturing
    b. Resource extraction
    c. Urbanization
    c. Long history of agriculture
    e. All of the above

4. At which stage of the Demographic Transition is Europe?
    a. First (high birth rate, high death rate)
    b. Second (high birth rate, declining death rate)
    c. Third (declining birth rate, declining death rate)
    d. Fourth (low birth rate, low death rate)
    e. Fifth stage (postindustrial, where population falls below replacement)

5. Recent migrants to England come mostly from all except one of the following countries. Which country is the exception?
    a. Hong Kong
    b. India
    c. Indonesia
    d. Jamaica
    e. Pakistan

6. Where are the "hard" borders of Europe?
    a. At the eastern edge of Europe, where the region meets Asia
    b. At the edge of continental Europe
    c. Between the countries of Europe
    d. Between Western and Eastern Europe
    e. On the perimeter of the European Union

7. All of the following statements about language in Europe are correct, EXCEPT…
   a. As their first tongue, 90 percent of Europe's people speak a Germanic, a Romance, or a Slavic language
   b. English is the official language of Europe
   c. Language has always been an important component of group identity in Europe
   d. Millions of Europeans learn multiple languages so they can communicate across cultural and linguistic boundaries
   e. Some small ethnic groups work hard to preserve their cultural identities and languages

8. What variation of Christianity is found in Europe?
   a. Roman Catholicism
   b. Orthodox Christianity
   c. Protestantism
   d. A and B above
   e. A, B, and C above

9. What is Europe's major geopolitical issue today, at the beginning of the 21st century?
   a. Integration of eastern and western Europe into the European Union
   b. The wars in Iraq and Afghanistan
   c. Region-wide environmental legislation
   d. Resolution of border disputes between European countries
   e. All of the above

10. Which of the following European countries was NOT part of the buffer zone for the former Soviet Union?
    a. Czechoslovakia   b. Sweden   c. Hungary   d. Poland   e. The Baltic States

11. Why did the Soviet Union build the Berlin Wall?
    a. To defend against attacks on its western boundary
    b. To serve as a routine border checkpoint
    c. To stop Germans from fleeing from East Germany to West Germany
    d. A and B above
    e. A, B, and C above

12. In which part of Europe has ethnic tension been so high for so long that its name has become a part of the word to describe the fragmented geopolitical process involved with small-scale independence movements and mini-nationalism as it develops along ethnic fault lines?
    a. Iberia   b. Scandinavia   c. The Balkans   d. The Baltic States   e. Benelux

13. When the first industrial districts appeared in continental Europe in the 1820s, what feature was the key determinant of location?
    a. Coalfields   b. Forests   c. Oil wells   d. Ports   e. Rivers and water wheels

14. Eastern Europe's sluggish economy and relatively poor economic performance are the result of its past history with which foreign country?
    a. China
    b. Japan
    c. Russia
    d. United States
    e. All of the above

15. All of the following statements about the economy of eastern Europe are true, EXCEPT...
    a. Latvia, Lithuania, and Estonia make up the middle range of economic and social change in the region
    b. Macedonia and Albania rank lowest in economic and social development in the region
    c. Problems of integrating the economies of eastern Europe into the region of Europe as a whole have recently resolved themselves
    d. Some countries of eastern Europe are speeding ahead to reap the rewards of successful change, while others appear to be falling even farther behind
    e. Successful economic transitions have occurred in the Czech Republic, Hungary, and Poland

Answers: 1-E; 2-A; 3-E; 4-E; 5-C; 6-E; 7-B; 8-E; 9-A; 10-B; 11-C; 12-C; 13-A; 14-C; 15-C

Chapter 9
THE RUSSIAN DOMAIN

LEARNING OBJECTIVES
- This chapter introduces the region that includes Russia and its neighboring states of Belarus, Ukraine, Georgia, and Armenia. This region has experienced an incredible degree of political upheaval since the 1990s and owes its existence in its current form to the breakup of the former Union of Soviet Socialist Republics (U.S.S.R.).
- The student should understand the role that the interaction between the United States and the U.S.S.R. has played in the 20th century in this region (especially after World War II).
- The student should also understand the problems posed by a cold, northern climate.
- In addition, the student should understand the differences between economic and political systems.
- Upon completion of this chapter, students should be familiar with the physical, demographic, cultural, political, and economic characteristics of the Russian Domain.
- Finally, students should understand the following concepts and models:

    · Cold War
    · Permafrost
    · Glasnost and perestroika
    · Russification
    · Denuclearization
    · Centralized economic planning

CHAPTER OUTLINE
I. Introduction
This region includes Russia, along with its neighbors Ukraine, Belarus, Georgia, Moldova, and Armenia, all formerly part of the Union of Soviet Socialist Republics (U.S.S.R,), which dissolved in 1991. The region has a population of about 200 million. Russia is the largest country on the planet and spans 11 time zones. This region is vast and resource rich, although it also includes some of the harshest climates found on Earth. In studying this chapter, students may find similarities between the rise of the United States and the rise of Russian culture; certainly the two countries have been engaged politically (and potentially militarily) in a Cold War for most of the last half of the 20th century. This region has also experienced extremely rapid political and economic change in recent years, moving uneasily from an authoritarian centrally planned economy (communism) and superpower status toward democracy and a capitalist economy. Currently, the region's economy is experiencing a boost from rising petroleum and natural gas prices, its commitment to democracy uncertain, and continuing nationalist movements in Russia threaten stability. Ukraine, Belarus, Georgia, and Armenia must all work to develop their own global relationships.

II. Environmental Geography: A Vast and Challenging Land (Fig. 9.4)

A. A Devastated Environment (Fig. 9.4)

1. This region has some of the world's most severe environmental degradation caused by industrialization, urbanization, careless resource extraction, and nuclear energy production; the problems here may have global impact

2. Air and water pollution; air pollution linked to clustering of industrial factories and minimal environmental controls; reliance on locally abundant, low-quality coal increases pollution; industrial pollution, raw sewage, oil spills, and seepage pollute water; pulp and paper factories built around Lake Baikal in the 1950s and 1960s polluted the lake; recent efforts have reduced pollution, but it's still a problem

3. The nuclear threat: the former Soviet Union (U.S.S.R.) had a large nuclear weapons and energy program, but environmental safety issues were ignored; nuclear weapons were used for seismic experiments, oil exploration, and dam building; nuclear wastes were carelessly dumped; aboveground nuclear testing left radioactive fallout in northeast Siberia; Russia has many aging nuclear reactors and has had two major nuclear accidents, one in the 1950s that contaminated hundreds of square miles and a 1986 meltdown in Chernobyl that devastated Belarus

4. The post-Soviet challenge: The end of Soviet control has had mixed results. The shutdown of non-viable factories reduced pollution, but brought economic decline. Advanced environmental equipment is beginning to be imported. Nuclear storage is being consolidated, environmental consciousness beginning to grow. The region has limited environmental regulations, and emphasis on quick profits on oil, natural gas, and timber worsens the situation.

5. Deforestation and poaching threaten wildlife

B. A Diverse Physical Setting: In this region, there are few domestic environmental regulations; as a result, the environment has suffered

1. This is one of the world's largest regions

2. It has many and varied natural resources

3. High latitude (northern) continental climate

4. Cold climate and rugged terrain limit opportunities for human settlement

C. The European West

1. European Russia, Belarus, Ukraine

2. Rivers (Dnieper, Don, Dvina, Volga) connected by canals and provide transport routes

3. Three environments influence agricultural potential in this region: (1) poor soils, cold temperatures, forests north of Moscow and St. Petersburg; (2) Belarus and central European Russia area has longer growing season, but acidic *podzol soils* limit farm output; (3) south of 50 degree latitude, grassland and fertile "black earth" *chernozem soils* support commercial production of wheat, corn, sugar beets, meat (Fig. 9.59)

D. The Ural Mountains and Siberia
1. Ural Mountains separate European Russian from Siberia; they are low mountains (under 1,000 feet), with a cold, dry climate that makes farming difficult; the Urals are heavily mineralized, and marked the eastern cultural boundary of Russia
2. Siberia lies to the east and extends for thousands of miles, has a very cold climate, with less than 20 inches of precipitation, mostly falling in summer; Siberia has three important rivers (Ob, Yenisey, Lena; vegetation includes *tundra* (with mosses, lichens, and ground-hugging flowering plants), and *taiga* (a coniferous forest zone, south of the tundra; eastern Siberia has *permafrost* (a cold-climate condition of unstable, seasonally frozen ground that limits growth of vegetation); farming is possible only in southwestern Siberia
E. The Russian Far East: found around Vladivostok (on the Pacific Ocean), is at about the same latitude as New England in North America; volcanic eruptions and earthquakes on the Kuril Islands and the Kamchatka Peninsula.
F. The Caucasus and Transcaucasia
1. Caucasus are found in the extreme southern portion of European Russia, some flat terrain, along with hills and the Caucasus Mountains; they form Russia's southern boundary; the highest peak is Mt. Elbrus (18,000 feet)
2. Georgia and Armenia are in Transcaucasia, which is dominated by low plateaus; the Lesser Caucasus Mountains are in

Southern Georgia and form the border between Armenia and Azerbaijan
3. Climate: high rainfall in west, eastern valleys arid or semi-arid, agriculture possible because of good soils
G. Consequences of Global Climate Change
1. Potential benefits include expansion of farmlands and more navigable waterways with warming.
2. Potential hazards include loss of cities (including St. Petersburg) with rising sea levels; disruption to indigenous peoples and wildlife in northern areas; thawing of permafrost could result in mudslides and release of frozen carbon materials.

III. **Population and Settlement: An Urban Domain**
A. About 200 million, widely dispersed throughout the region
B. Population Distribution (Fig. 9.13)
1. Population distribution: there are more people in the best agricultural regions (in the west); European Russia has more than 100 million, while Siberia has only 35 million; 60 million in Belarus, Moldova, and Ukraine
2. The European Core (Belarus, much of the Ukraine, western Russia): has region's largest cities (Moscow, Nizhney, Novgorod), biggest industrial centers, best farms; St. Petersburg was the capital of the Russian Empire
3. Siberian hinterlands: relatively sparse settlement, with two zones: (1) industrial cities along the south side of the Trans-Siberian Railroad completed in 1904; (2)

thinner settlements along the Baikal-Amur Mainline Railroad (BAMRR), north of Trans-Siberian completed in 1984; north of BAMRR, few settlements exist

C. Regional Migration Patterns: three waves of migration (Fig. 9.16)
1. Eastward movement (1860–1914) accelerated by Trans-Siberian Railroad; settlers attracted by farming opportunities in the south and greater political freedom away from the *tsars* (czars); almost 1 million settlers to Siberia from 1860–1914; continued under Soviet regime
2. Political Motives: moves dictated by Russian leaders: (1) movement to Siberia to exploit natural resources; (2) millions forcibly relocated to political prisons in Siberia (*Gulag Archipelago:* a vast collection of political prisons in which inmates often disappeared or spent many years removed from home and family); (3) *Russification:* Soviet policy of resettling Russians into non-Russian portions of U.S.S.R to increase Russian dominance; Russians became a significant minority in former Soviet republics and often received special treatment
3. New International Movements: reversal of Russification after breakup of U.S.S.R., several formerly "Russified" republics imposed rigid citizenship and language requirements to encourage Russians to leave; more open borders have contributed to "brain drain" of young, well-educated, upwardly mobile Russians; Jews moving to Israel or United States

D. The Urban Attraction
1. Marxist philosophy of Soviet planners encouraged urbanization
2. Soviets carefully planned cities; selecting cities for specialized purposes, with predetermined population levels; internal passports kept people from moving freely as people worked in the cities where the government sent them
3. Since breakup of the U.S.S.R., Russian citizens have greater freedom of movement; many older industrial areas are now losing population
4. Inside the Russian city: carefully planned, including a center with superior transit connections, department stores and shops, housing and offices; inner-city decay is rare, no suburban sprawl; concentric rings: (1) public housing and *sotzgorods* (socialist neighborhoods based on a close connection between workplace and home, often dormitories); (2) the *Chermoyuski:* large, uniform apartment blocks shoddily built in the 1950s/1960s with the idea that they would be replaced after economic growth; (3) *mikrorayons:* much larger housing projects of 1970s/1980s, self-contained community with stores and services nearby; (9–24 stories) about 45 minutes from city core; (4) rural *dachas* (country houses) available only to the elite

E. The Demographic Crisis
1. General population decline in Russia, Belarus, Ukraine caused by low birth rates and rising death (mortality) rates, especially among middle-aged males

2. Causes: fraying social fabric; uncertain economy; declining health of women of child-bearing age; stress-related diseases (alcoholism and heart disease); rising murder and suicide rates, toxic environments

3. Recent government incentives to encourage more births include cash payments for additional children in a family, extended maternity leave and day care subsidies; birth rates have recently started to rise slightly.

IV. Cultural Coherence and Diversity: The Legacy of Slavic Dominance

A. Slavic, Russian-speaking peoples diffused from central European Russia

B. Heritage of the Russian Empire
1. Russian expansion paralleled European colonization: in 1500s and 1600s Russia expanded eastward, advancing into Central Asia, to the border of Korea and Sea of Japan; the empire remained intact till 1991 (long after European countries gave up their colonies)

2. Origins of the Russian state: Slavs originated near Pripyat marshes (Belarus), began eastward migration about 2000 years ago; migration to the west gave rise to Poles, Czechs, and Croations; intermarriage with warriors from Sweden (known as *Varangians* or *Rus*), led to the name "Russian"; interaction with Byzantine Empire of Greece led to adoption of *Eastern Orthodox Christianity* (a form of Christianity linked to eastern Europe and Constantinople) and the Cyrillic alphabet; Poles, Czechs, Slovaks,

Slovenians, and Croations adopted Catholicism

3. Growth of the Russian Empire (rapid eastward expansion): Slavic leaders in Moscow overthrew Tatars in late 1400s, established new, unified Russian state, and Russia became a nationalistic empire with a divine mandate to spread as far as possible; Russians allied with *Cossacks* (semi-nomadic, Slavic-speaking Christians), who conquered Siberia; Japanese stopped further expansion (Russo-Japanese War of 1905)

4. Growth of the empire: westward expansion was slow and halting, Russia gained territory from Sweden, Poland, and Ottoman Turks by the 1700s, led to creation of St. Petersburg, and addition of Belarus and Ukraine; final expansion occurred in 1800s, adding Georgia and Armenia

5. The legacy of empire: tightly integrated Russian culture from St. Petersburg to the Sea of Japan; but many pockets of linguistic, national and religious differences and tensions remained; Russia and its domain maintained an ambivalent relationship with the West, attracted by the modern technology but repelled by European culture and social institutions; Russian governments have a long history of authoritarianism

C. Geographies of Language (Fig. 9.21)
1. Slavic languages dominate
2. Patterns in Belarus, Ukraine, and Moldova: Belarus is more or less a nation-state with scattered Polish and Russian minorities; Ukraine includes speakers of

Russian in eastern Ukraine and Crimea; some Ukrainians live in southern Russia and southwestern Siberia; Moldavians speak a dialect of Romanian (a Romance language).

3. Patterns Within Russia: ethnic Russians dominate, but there are pockets of indigenous peoples; some are seeking autonomy; non-Russians include Finno-Ugric and Uralic-Altaic groups, and Eskimo-Aleuts in Siberia

4. Transcaucasian languages: north slope of Caucasus, to Georgia and Armenia, with one of the world's most complicated language patterns: three language families in an area smaller than Ohio, several dozen individual languages remain; language is a pivotal cultural and political issue in Transcausasian area

D. Geographies of Religion

1. U.S.S.R. prohibited open practice of religion, and converted most houses of worship to other uses; during WWII, many people practiced religion privately; demise of U.S.S.R. brought religion back into the open

2. Contemporary Christianity: Most Russians, Belorussians, Ukrainians are followers of Eastern Orthodox Christianity In the Caucasus, Armenian Christianity prevails and differs from both Eastern Orthodox and Catholic traditions; Georgian Christianity similar to Orthodox; United States missionaries bringing evangelical Protestantism since demise of U.S.S.R.; Roman Catholicism is found in Western Ukraine.

3. Non-Christian religions: 20–25 million Sunni Muslims live in the North Caucasus, including Chechnya; about 20% of Moscow's people are Muslim; more than 1 million Jews live in Russia, Belarus, and Ukraine, especially in larger cities of the European West; many Jews have emigrated to escape persecution

E. Russian Culture in Global Context

1. Russian culture has strong inward orientation, most common people having little interaction with outside world; Russian high culture is westernized, Russian composers, novelists, playwrights famous in Europe and United States (Tchaikovsky, Dostoevsky, Chechov)

2. Soviet days: European-style modern art flourished at first, but in late 1920s, leaders scorned modernism as a "decadent expression of a declining capitalist world"; many artists fled or were exiled to Siberian labor camps; Soviets promoted *"socialist realism,"* a style devoted to the realistic depiction of workers harnessing the forces of nature or struggling against capitalism; traditional high arts (classical music, ballet) with no obvious political connotation received lavish state subsidies

3. Turn to the West: in the 1980s, young people adopted rebellious approach and United States-style mass consumer culture popular; after the fall of the U.S.S.R., Western products like rock music, chewing gum, jeans, and other consumer goods and films came to Russia from the West and elsewhere (India, Hong Kong,

Latin America); censorship became more difficult

4. Russians and the Net: Internet use tripled to about 7 million, or about 5 percent of the people in the region between 1999 and 2003; this could have longer-term economic and cultural impacts

5. The music scene: American and European popular music is gaining fans, and a homegrown music industry is evolving; Russian MTV reached more than 60 million viewers by 2001; Sony Music established Russian operations in late 1999; music piracy is widespread

6. Revival of Russian nationalism: government and the media in Russia are reviving Russian symbols to cultivate a new sense of Russian identity in the region (national anthem, holidays, celebrations, awards)

V. **Geopolitical Framework: The Remnants of a Global Superpower** (Fig. 9.30)

A. Many of today's problems in this region stem from the Soviet legacy

B. Geopolitical Structure of the Former Soviet Union

1. Soviet Union arose after Russian Empire collapsed abruptly in 1917; authoritarian and aristocratic tsars were replaced by a broad-based coalition of business people, workers, and peasants, who were replaced a few months later by the *Bolsheviks* (a faction of Russian Communists representing the interests of the industrial workers), led by Lenin (Vladimir Ilyitch Ulyanov), who centralized

power and introduced communism

C. The Soviet Republics and Autonomous Areas

1. Leaders designed a geopolitical solution to maintain the country's territorial boundaries and theoretically acknowledged the rights of non-Russian citizens

2. Each major nationality outside the traditional boundaries of Russia received its own *"union republic"* (Fig. 9.230); this created 15 republics, each one administratively autonomous, with the theoretical right to secede; but the U.S.S.R. remained a centralized state

3. Inside Russia's boundaries, the U.S.S.R. created *autonomous areas:* ethnic homelands of various sizes within the structure of the U.S.S.R

D. Centralization and Expansion of the Soviet Union

1. The policies associated with autonomous regions did not cause ethnic differences to disappear, so in 1930, Soviet leader Joseph Stalin centralized power in Moscow to assert Russian authority; national autonomy disappeared; Stalin introduced state-controlled farms and industries and used force against citizens to make them comply— especially anti-Soviet ethnic groups; Stalin enlarged the U.S.S.R., adding Sakhalin and Kuril Islands from Japan, regained Baltic republics (which were independent from 1917 to 1940), and occupied parts of Poland, Romania and Czechoslovakia; also added a small portion of Germany, which is still a Russian

exclave (a portion of a country's territory outside of its contiguous land area)

2. After World War II: U.S.S.R. gained authority (but not sovereignty) over much of eastern Europe; established an *"Iron Curtain"* between their eastern European allies and the more democratic nations of western Europe; U.S.S.R. sent troops to Hungary and Czechoslovakia; 1980s saw decline of Soviet authority, undermining communism

3. The *Cold War* (1948–1991): Soviet Union and United States became political and military rivals; U.S.S.R. formed alliances with countries around the world (e.g., North Korea, Vietnam, Cuba, Angola, Somalia; and fought a war with Afghanistan from 1979 to the late 1980s— U.S.S.R. lost

4. End of the Soviet System: Lenin's culturally defined republics provided persisting political framework for ethnic nationalism that grew after WWII; in 1980s, Mikhail Gorbachev instituted *glasnost* (greater political openness); *union republics* (especially the Baltic republics of Latvia, Lithuania, and Estonia) began to demand independence; other forces led to demise of U.S.S.R.: (1) worsening economic conditions; (2) introduction of *perestroika:* (planned economic restructuring to make production more efficient and more responsive to needs of Soviet citizens); by the end of 1991, all of the country's 15 union republics became independent states

E. Current Geopolitical Setting (Fig. 9.31)

1. Russia and the former Soviet republics formed Commonwealth of Independent States (CIS), but it is mostly a forum for discussion, with no economic or political power; Russia is making ties with Belarus and Ukraine; Georgia, Ukraine, Azerbaijan, Moldova have organized as a group focused on trade (especially oil); *denuclearization* (return and partial dismantling of nuclear weapons from outlying republics to Russian control) was completed in 1990s; military, political, and ethnic tensions remain in parts of the region

2. Geopolitics in the South and West: Ukrainians favored reformist candidates, as Russia tries to influence its politics; Moldova, Georgia, Armenia also experience instability

3. Geopolitics within the Russian Federation: *devolution:* more localized political control; the Russian Federation Treaty of 1992 grants greater political, economic, and cultural freedoms to internal units; in some places, this has encouraged these units to pressure Russia for even more autonomy; Russia has renewed its efforts to assert its national presence

4. Russian challenges to civil liberties: President Putin consolidated political power, and reduced democracy; it is unclear what will happen with the election of President Medvedev (supported by Putin) in 2008

F. A growing global presence: Russia is reasserting its political status, and is a member of the G-8 (an organization of economically powerful countries in the world)

VI. Economic and Social Development: An Era of Ongoing Adjustment

A. After GNP decline in the 1990s, Russia's economy has been stable since 2002
1. Causes: policies of the Soviet (communist) era; the chaotic nature of economic reforms and their uneven applications in the 1990s
2. Rising demand for oil and natural gas has helped to improve Russia's economy, but problems remain

B. Legacy of the Soviet Economy
1. The Communists came to power in 1917 in a quick, unexpected revolution, and instituted *centralized economic planning:* a situation in which the state controls production targets and industrial output (communism)
2. By the 1920s, most of the economy was controlled by the Soviet state; many factories were virtual slave-labor camps, while family farms were combined into large-scale collective farms *(kolkhozes)*; resulting famines caused many to starve
3. Soviet industry was more successful than its agriculture; Soviets added major industrial zones (Fig. 9.31), many near energy sources and metals; Moscow had fewer raw materials, but had some of Russia's best infrastructure, large pool of skilled labor, and demand for industrial products

4. Soviets developed a good transportation and communication infrastructure, had a massive housing campaign in 1960s, made literacy virtually universal and health care readily available; eliminated the worst of the poverty
5. During the 1970s and 1980s, economic and social problems increased, agriculture did not cover all of U.S.S.R.'s needs; manufacturing efficiency and quality lagged behind the West; shortages occurred and people waited in lines for basic goods; U.S.S.R. did not participate fully in the technological revolutions that transformed the United States, Europe, and Japan; there were growing disparities between elites and average people, and few gains in personal and political freedom

C. The Post-Soviet Economy
1. Since the demise of the U.S.S.R., the region has dismantled much of its centralized, state-controlled economy, replacing it with a mix of state-run operations and private enterprise; it has been a difficult transition
2. Redefining regional economic ties: since the demise of the former U.S.S.R., the now-independent republics must negotiate for needed resources with each other, rather than accept centralized exchanges; this increases the cost of doing business; Russia continues to dominate region's economy
3. Privatization and state control: Russia removed price controls in 1992, causing prices to rise and encouraging inflation; wages did

not keep pace, the value of the Russian ruble dropped; in 1993, the Russian government allowed citizens to buy into newly privatized farmlands and industries; lack of controls allowed corruption and mismanagement in the new system; other republics lagged behind Russia in privatizing; by 1995, 70 percent of Russia's wealth came from private enterprise; some elite members of Russian society have become wealthy through these changes; most workers have seen little economic improvement

4. The challenge of corruption: organized crime pervades Russia and controls many parts of the economy; Russia's interior ministry estimates that the Russian mafia controls 40 percent of the private economy and 60 percent of state-run operations; there are ties between the Russian mafia and Russian intelligence; the Russian mafia has gone global and was implicated in a major money-laundering scheme involving Russian, British, and United States banks; the Russian mafia has invested in every corner of the world

D. Ongoing Social Problems
1. Uncertain economy and politics affect social life
2. Problems: (1) organized crime; (2) rising unemployment; (3) higher housing costs; (4) declining social expenditures; (5) increased domestic violence; (6) rising divorce rates; (7) reduced government spending on education; (8) reduced health-care spending resulting in vaccine shortages and other problems; (9) increase in stress- and environmental-related illnesses

E. Globalization and Russia's petroleum economy:
1. Russia has 35 percent of the world's natural gas reserves (mostly in Siberia) and is the world's largest gas exporter; Russia is the largest non-OPEC oil exporter; the primary destination for Russian petroleum products is western Europe; construction of new pipelines, which United States, European, and Japanese interests are supporting, will improve the transport of Russian petroleum products
2. Local impacts of globalization: local impacts of globalization are very selective; large cities are more likely to have imported goods available than smaller cities; cities that produce products for sale on the world market also see greater impacts of globalization; port cities are well-positioned to take advantage of their accessibility to nearby markets; globalization has made it difficult for older, less-competitive industrial centers to find a market for their goods, in large part because their days as part of the U.S.S.R. left them poorly prepared for competing in a global market of consumer goods; reductions in prices for commodities, such as extractive minerals, mean the region is vulnerable to fluctuations in the market prices for these products

F. Growing Economic Globalization

1. Entry in world market began in 1970s, when U.S.S.R. began to export oil and import food
2. A new day for the consumer: new consumer imports now come to the region, especially to larger cities; these include McDonald's, Calvin Klein jeans; luxuries like BMWs are available, but out of the reach of most Russians; other imports, such as frozen chickens from the United States, inexpensive consumer goods from East and Southeast Asia and from industrializing neighbors like Turkey, are also available
3. Attracting foreign investment: the Russian Domain struggles to attract foreign investment; the United States and western Europe (especially Germany and Great Britain) the strongest investors; most investment is in oil, gas, food, telecommunications, and consumer goods industries; many outsiders are wary of investing in the region because of the unstable economy, the mafia, and government red tape

VII. Conclusions

The Russian Domain is a region that has seen great change, beginning with its rise as an empire, through revolution and breakup. Ethnic and cultural differences have long played a role in this region and continue to shape its political destiny. Although this region is rich in natural resources, its limited agricultural potential contributes to lingering economic difficulties. As well, the massive readjustments growing from the political and economic upheavals of the 1990s continue to affect the area. Even if these political and economic upheavals were resolved quickly, the region must still face the environmental devastation it has experienced. The effects on the people have been dramatic, contributing to social and health problems as well. Much uncertainty lies ahead for the people of the Russian Domain.

PRACTICE MULTIPLE CHOICE QUIZ

1. In addition to Russia, which other countries are part of the Russian Domain?
   a. Ukraine and Belarus
   b. Georgia and Armenia
   c. Moldova
   d. A and B above
   e. A, B, and C above

2. All of the following statements about the environment in the Russian domain are true, EXCEPT...
   a. The environment has been vastly improved since countries in the region achieved independence in the 1990s
   b. The environment was degraded by careless development under the Soviet Union
   c. Municipal water supplies in the region are vulnerable to industrial pollution and raw sewage
   d. Nuclear pollution is pronounced in northern Ukraine
   e. Poor air quality plagues hundreds of cities in the region

3. What is the key to understanding the basic geographies of climate, vegetation, and agriculture in the Russian domain?
   a. The region's altitude
   b. The region's northern latitude
   c. The region's topography
   d. The region's waterways
   e. The region's seismic activity

4. What is the significance of the Ural Mountains in the Russian Domain?
   a. They rival the Himalayas in height
   b. They are the oldest mountains in the world
   c. They are the site of major development in the region
   d. They are the widest mountains in Eurasia
   e. They separate European Russia from Asian Russia

5. What is the major reason for the strikingly higher population densities in the European West part of the Russian Domain?
   a. The nearness of this region to Europe
   b. The people of this region have a culture that encourages large families
   c. The Soviet Union established this pattern with its settlement policies
   d. This is where the best agricultural lands in the region are found
   e. This region attracts many immigrants from outside the Russian Domain

6. All of the following statements about Siberia are correct, EXCEPT…
   a. Populating Siberia had a political motive
   b. Siberia is located in the western part of Russia
   c. Subarctic climates are found in Siberia
   d. The Gulag Archipelago, a vast collection of political prisons, was located in Siberia
   e. Siberia became a repository for political dissidents in the Soviet Union

7. People from which neighboring country are crossing the Amur River border (sometimes illegally) into Russia to trade, work, and live?
   a. China   b. Finland   c. Kazakhstan   d. Lithuania   e. Mongolia

8. What has been the major factor in recent migration in the Russian Domain?
   a. Deteriorating environmental conditions in some parts of Russia
   b. Extensive foreign direct investment in the Donetsk are, resulting in a booming economy there
   c. Freedom of mobility arising from the breakup of the former U.S.S.R.
   d. Internal ethnic conflicts have created many internally displaced persons
   e. Russian policies assigning people to specific cities for work and residency

9. What is the status of the population in Russia?
   a. After a short decline immediately following the breakup of the U.S.S.R., the population is now growing rapidly

b. The population has seen a steady increase since the breakup of the U.S.S.R.
c. The population is holding steady
d. The population is in a state of persistent decline
e. There are no recent population statistics available to answer this question

10. What is the dominant language group in the Russian Domain?
    a. Altaic   b. Romance   c. Finno-Ugric   d. Slavic   e. Germanic

11. The Soviet Union's invasion of this country in 1978, and the long war that followed, helped to bring about the end of the Soviet era.
    a. China   b. Afghanistan   c. Mongolia   d. Poland   e. Turkey

12. What is the status of the Commonwealth of Independent States (CIS)?
    a. It has become an important international trade organization
    b. It has disbanded
    c. It continues to exist, but it has no real economic or political power
    d. It is actively following the path established by the European Union
    e. It is growing in importance as a body to handle political and ethnic conflicts with the region

13. What region of Russia tried to secure its independence from Russia, only to have Russia move in a large number of troops to reassert its control?
    a. Chechnya   b. Siberia   c. Mordvinia   d. Sakha   e. Tatarstan

14. Which of the following is an undisputed economic advantage of the Russian Domain?
    a. Abundant natural resources
    b. A well-educated population
    c. Vast size
    d. Urbanization
    e. All of the above

15. What provides the strongest global link between the Russian Domain and the rest of the world?
    a. The region's agricultural exports
    b. The region's cultural strengths
    c. The region's growing software industry
    d. The region's manufacturing sector
    e. The region's oil and gas industry

Answers: 1-E; 2-A; 3-B; 4-E; 5-D; 6-B; 7-A; 8-C; 9-D; 10-D; 11-B; 12-C; 13-A; 14-E; 15-E

Chapter 10
CENTRAL ASIA

LEARNING OBJECTIVES
- This chapter introduces Central Asia, which includes Mongolia, Afghanistan, and six countries that were part of the former Soviet Union (Azerbaijan, Kazakhstan, Kyrgyzstan, Uzbekistan, Tajikistan, and Turkmenistan).
- This region also includes the world's tallest mountains, the Himalayas.
- This region has become more familiar since September 11, 2001, when it became apparent that the attack on the World Trade Center had been planned and organized by Osama bin Laden and his al-Qaeda organization, which had been based in Afghanistan (even though bin Laden is a Saudi national, and al-Qaeda operates globally).
- The students should understand the significance of this region's landlocked location, the only geographical region described in the text that is lacking ocean access.
- Students should also understand the historical cohesion of this region, along with its pivotal role in the evolution of the Eurasia.
- In addition, students should understand the effect of continentality and terrain on climate patterns.
- Upon completion of this chapter, students should be familiar with the physical, demographic, cultural, political, and economic characteristics of Central Asia.
- Student should also understand the following concepts and models:

    · Continental climate
    · Exotic river
    · Desiccation
    · Buddhism
    · Landlocked areas
    · Turkestan

CHAPTER OUTLINE

I. Introduction

This region covers the central part of the Asian continent and includes the countries of Mongolia and Afghanistan, along with six former republics of the now-defunct U.S.S.R. Historically, this region has played a significant role in the evolution of Eurasia as a whole. As well, it is the only world region that has no access to an ocean; it is landlocked.

The physical geography of this region is remarkable because of the presence of the Himalayas and other mountain chains as well as several deserts. Because of the rugged terrain and dry climate, population density in this region is low. While the environmental unity of the region is apparent, the cultural differences among the region's residents are great. Oil is an important natural resource in this region. Within recent memory, either the former U.S.S.R. or

China has dominated virtually this entire region. The United States currently has troops stationed in Afghanistan.

II. Environmental Geography: Steppes, Deserts, and Threatened Lakes of the Eurasian Heartland

A. The Shrinking Aral Sea (world's 4th largest lake), located on the border of Kazakhstan and Uzbekistan (Fig. 10.2.1)
1. The Aral Sea is not a true sea, because it is not connected to an ocean; its sources of water are two rivers that are heavily used for irrigation; as more water was diverted for crops, less made it to the Aral Sea; salinity increased, causing most fish to die
2. The destruction of the Aral Sea has led to economic and cultural damage and ecological devastation; some formerly "seaside" villages are now as many as 40 miles from the Aral Sea; fisheries have closed

B. Other Environmental Issues (Fig. 10.3)
1. Much of Central Asia has a relatively clean environment, mostly because of the low population density; industrial pollution is a problem only in the region's largest cities; other problems in the region are typical of arid regions: (1) desertification, (2) salinization, (3) desiccation (drying up of lakes and marshes)
2. Shrinking and Expanding Lakes: the Caspian Sea is the world's largest lake, about the size of Montana (like the Aral, it isn't connected to an ocean, so it isn't a true sea); its water sources are the Ural and Volga Rivers; dams and irrigation divert water from the Caspian; its level dropped to a low point in the 1970s; today it is eight feet higher than that; pollution from industry, especially from oil drilling, is a problem
3. *Desertification* (spread of deserts): in the eastern part of the region, the Gobi Desert has spread southward into China; China has tried (with some success) to reverse the trend by massive efforts to plant trees and grass
4. Deforestation: most of Central Asia is too dry to support forests, but many of its mountains were once well wooded
5. *Salinization:* accumulation of salt in the soil

C. Central Asia's Physical Regions (Fig. 10.1)
1. In general Central Asia has steppes (grassland plains) in the north, desert basins in the southwest and central areas, and high plateaus and mountains in the south-center and southeast
2. The Central Asian Highlands: originated with collision of Indian subcontinent into Asian mainland, creating the Himalayan Mountains (highest in the world) and the Karakorum, Pamir, Hindu Kush, Kunlun Shan, and Tien Shan ranges; Tibetan Plateau stretches 1,250 miles east-west, 750 miles north-south with an average height of 15,000 feet, near the maximum elevation at which human life can exist; most rivers of South, Southeast, and East Asia (Indus, Ganges, Brahmaputra, Salween, Mekong, Yangtze, and Huang He) originate in the Tibetan

Plateau and nearby mountains; most of Tibet is arid, with cold winters, and warm summers with chilly nights

3. The Plains and Basins: west of the Tien Shan and Pamir Mountains are the Kara Kum (black sand) and Kyzyl Kum (red sand) deserts; the Aral Sea is 135 feet above sea level; Caspian is 92 feet BELOW sea level; this area has a continental climate (Fig. 10.5), with hot, dry summers and winters averaging well below freezing temperatures; temperatures more moderate along the Caspian; the region's eastern desert belt (Taklamakan) stretches 2,000 miles from western China to Inner Mongolia; rainfall increases toward the north of this area, as desert becomes steppe

D. Global Warming and Central Asia
1. Reduction of permafrost and loss of glaciers is already occurring, reducing river flows
2. The Gobi Desert and the Tibetan Plateau could see more precipitation.

III. Population and Settlement: Densely Settled Oases Amid Vacant Lands

A. Most of the region is sparsely populated (Fig. 10.6)
1. Many places are too arid or too high to support human life
2. Many of the more favorable areas are populated by *nomadic pastoralists* (people who raise livestock for subsistence purposes); sedentary people live in the river valleys

B. Highland Population and Subsistence Patterns

1. Harsh environment: mountain tundra has cold temperatures, scarce water, pronounced ultraviolet radiation because of high elevation
2. Most of the Tibetan Plateau supports only *nomadic pastoralism;* peoples here created empires based on their military power; extremely low population densities; in the Pamir Range, there are isolated valleys that support agriculture and intensive human settlement

C. Lowland Population and Subsistence Patterns:
1. Most desert dwellers live where the mountains meet the basins and plains, there is adequate water, and there is no salt or alkali residue in the soils; intensive cultivation occurs amid the *alluvial fans* (fan-shaped deposit of sediments dropped by streams flowing out of mountains) in the area; the silty *loess* soils deposited by the winds in the area are fertile
2. The Gobi Desert is one of Asia's least populated areas; steppes of northern Central Asia support nomadic pastoralism, but some of the area has been planted in wheat

D. Population Issues
1. Central Asia is a low-density environment, but some portions are growing at a moderately rapid pace
2. Han Chinese have been immigrating to Central Asia, while ethnic Russians have been emigrating from Kazakstan
3. Migration from the region to Moscow from former Soviet republics

4. Oil is attracting migrants to the region
    E. Urbanization in Central Asia
        1. Older cities (i.e., Samarkand and Bukhara) known for lavish architecture, and located on international overland trade routes; when sea trade replaced overland trade, these cities fell on hard times
        2. Cities' construction under Russian and Chinese control reflect those more austere styles
        3. Overall the region remains largely rural, with fewer than one-third living in cities

IV. **Cultural Coherence and Diversity: A Meeting Ground of Different Traditions**
    A. Historical Overview: Steppe nomads and Silk Road traders
        1. Farming villages dating from the neolithic era (8000 BCE) have been discovered; domestication of the horse made nomadic pastoralism possible; this livelihood dominated because the mobility offered protection from enemies; the invention of guns offset the advantages of nomadic pastoralism
        2. Early languages in Central Asia are from the Indo-European family, related to Persian; this region is considered the birthplace of Indo-European peoples; Altaic languages (Tungusic, Mongolian, Turkish) replaced Indo-European about 200 years ago; Turks spread from what is now Mongolia, then Mongols replaced Turks; by late 1200s, Mongol Empire was the largest the world had seen

3. Tibet had a strong, unified kingdom by A.D. 700, but its independence has been elusive
    B. Contemporary Linguistic and Ethnic Geography
        1. Most peoples of Central Asia speak Mongolian or Turkic languages (Fig. 10.13); a few indigenous Indo-European languages are found in the southwest, while Tibetan is the main language of the plateau; Russian is found in the west, Chinese is growing in importance in the east
        2. Tibetan: although it's part of the Sino-Tibetan language group, the linkage between Chinese and Tibetan is unclear; about 1.5 of 2.5 million people in Tibet speak Tibetan, and the other 1 million speak Chinese; Tibetan has an extensive literature, most of it on religious topics
        3. Mongolian: about 5 million speak Mongolian; 90 percent of people in Mongolia speak the language; in 1941, Mongolia adopted the Cyrillic script, but there are current efforts to revive the old script
        4. Turkic languages are the most common in the region—more than Tibetan and Mongolian combined; spoken from Azerbaijan (west) to China's Xinjiang province (east); Uygurs comprised 90% of Xinjiang's population in 1953, but immigration by Han Chinese has caused the Uygurs to be a minority in their own homeland now; other Turkic languages are Azeri, Uzbek, Turkmen, Kazak, Kyrgyz

5. Linguistic Complexity in Tajikistan: Tajik is related to Persian; Azerbaijan also has a complex language mix that includes Turkish; Soviet policy supporting ethnic homogenization increased the complexity of language in the region

C. Language and Ethnicity in Afghanistan (Fig. 10.15)
   1. Afghanistan was never colonized; its boundaries are those of its original, premodern, indigenous kingdom that emerged in the 1700s
   2. Two major language groups in Afghanistan: Pashtun (40–60 percent), most of whom live along the Pakistani border, and Dari (related to Persian), mostly living in western Afghan cities, central mountains, and near the Tajikistan border; another 11 percent speak Turkic languages
   3. The ethnic diversity was not a problem for most of Afghanistan's history, but came out in last part of 20th century

D. Geography of Religion
   1. Because of this region's location along early overland trade routes, many religions exist here: Buddhism, Islam, Christianity, Judaism, Zoroastrianism, and several minor religions
   2. Islam in Central Asia: various groups interpret Islam differently; Pashtuns of Afghanistan interpret Islam extremely strictly, based on their pre-Islamic customs; nomadic Kazaks interpret Islam more loosely; most Muslims are Sunni, but Shi'ism is dominant among Hazeras (of central Afghanistan) and the Azeris; both Russian and Chinese communists tried to suppress religions, including Islam; interest in Islam is increasing as people reassert their indigenous identity; when the Taliban controlled Afghanistan, they insisted that all aspects of society conform to its harsh version of Islamic orthodoxy; in 2001, the Taliban destroyed all Buddhist statues in the country, including the largest works of art; the Taliban was removed from power in 2001
   3. Tibetan (Lamaist) Buddhism: Mongolia and Tibet are mostly Buddhist; before the Chinese dominated Tibet, the country was essentially a *theocracy* (religious state), with the Dalai Lama enjoying political and religious authority; Lamaism is hierarchically organized and committed to monasticism; Chinese invaded Tibet in 1959, and persecuted Lamaist Buddhism since, including destruction of 6,000 monasteries and killing of thousands of monks; the Dalai Lama lives in exile in India; Lamaist Buddhism is experiencing a resurgence in Mongolia

E. Central Asian Culture in International and Global Context
   1. Western Central Asia's closest contact is Russia; Russian language and Cyrillic alphabet spread throughout the region; demise of U.S.S.R. meant decline of Russian influence
   2. Eastern Central Asia's closest contact is China
   3. United States culture is growing; use of English is growing,

especially among people with computer skills

V. Geopolitical Framework: Political Reawakening in a Power Void (Fig. 10.18)

A. Central Asia has played a small role in global political affairs for several hundred years, although in recent years, Russia and China have figured most prominently

B. Partitioning of the Steppes
1. Before 1500 and the invention of guns, Central Asia was a major power center, with mobile horseback armies
2. Manchu Dynasty of China occupied Mongolia, Xinjiang, Tibet, and part of Kazakstan by mid-1700s (Mongolia and Tibet declared independence in the 1900s)
3. Russian Empire began conquests in this region in 1700s, finished just before the Soviet era (1917–1991)

C. Central Asia under Communist Rule
1. Soviet Central Asia: Soviets imposed socialist economy in the region and encouraged Russian immigration, forced indigenous peoples to write their languages in Arabic script; Soviets also established *union republics,* which had a certain level of autonomy (see Chapter 9) in an effort to encourage indigenous peoples to develop a Soviet identity, instead they encouraged nationalism that eventually led to breakup of the U.S.S.R.
2. The Chinese Geopolitical Order: after emerging as a communist country in 1949, China regained its dominance in Central Asia; China promised regional autonomy to Xinjiang to gain its support; Tibet did not want to give up its independence; Chinese detailed a policy similar to Russian union republics that was not carried out in practice; China encouraged immigration of Han Chinese to the region— this caused additional problems

D. Current Geopolitical Tension
1. Former Soviet union republics made mostly smooth transition after 1991, though some ethnic tensions remain; non-Han Chinese peoples are unhappy with Chinese domination
2. Independence in former Soviet lands: all six Central Asian union republics joined the Commonwealth of Independent States; democracy has made less progress in Central Asia than in other parts of the former U.S.S.R.; ethnic tensions remain in Kazakstan and Tajikistan; Armenia invaded Azerbaijan after breakup of U.S.S.R
3. Strife in western China: local opposition to Chinese rule grew in the 1990s; Chinese military occupies Tibet (which has valuable mineral deposits and is the site of nuclear weapons tests) and represses protest there; Uygur separatist movement is evolving; China believes all of its possessions in Central Asia are integral to its national territory
4. War in Afghanistan before 9/11/2001: began in 1978, when Soviet-supported military revolutionary council seized power, leading to rebellion (mujahedeen, who were armed by Pakistan, Saudi Arabia, and the United States); Soviets

withdrew in 1989, their puppet government endured for several years; in 1995–1996, the Taliban came to power through conquest and installed a repressive Islamist government; by late 1990s, most Afghans disillusioned with Taliban and its severe restrictions on everyday life (especially for women)

E. Global Dimensions of Central Asian Geopolitics
 1. Several countries vie for power and influence in Central Asia: China, Russia, Pakistan, Iran, Turkey, and the United States
 2. Revival of Islam in the region has also generated international geopolitical impacts
 3. A Continuing United States Role: U.S. military in Afghanistan, Uzbekistan, Tajikistan, Kyrgyzstan; Russian military base in Kyrgyzstan.
 4. Relations with China and Russia: boundary between China and Tajikistan and Kyrgyzstan resolved; Russia has interest in former republics, including pipeline construction; Shanghai Cooperation Organization (SCO) or Shanghai 6: China, Russia, Kazakhstan, Kyrgyzstan, Tajikistan, Uzbekistan are cooperating as a counterbalance to the United States
 5. International relations within Central Asia: tensions with China regarding Muslim Uyghurs; Uzbekistan has mined its borders with Tajikistan and Turkmenistan
 6. The Roles of Russia, Iran, Pakistan, and Turkey: Russia continues to have an interest in Central Asia, especially in the countries that were former republics of the U.S.S.R.; Iran is a major trading partner in this region and has strong cultural ties (although religious ties are weaker because Iran's Muslims are mostly Shi'a); Pakistan would like to increase its influence and had been a strong supporter of the Taliban before September 11, 2001; Turkey has close cultural and linguistic ties with the region, and Turkey offers itself as an alternative model of a modern Muslim country; a pan-Turkic (Turkestan) movement is possible

VI. Economic and Social Development: Abundant Resources, Devastated Economies

A. Economic Development in Central Asia: Central Asia is one of the least prosperous regions in the world
B. The Postcommunist Economies
 1. None of the countries in this region is considered prosperous
 2. Kazakhstan has good agricultural potential, a small population, and could become a food exporter; has deposits of oil and natural gas, and is exploring prospects for export; has signed agreements with Western oil companies, but red tape is slowing progress
 3. Uzbekistan has region's second-largest economy, and its economy has not declined as sharply as those of its neighbors, partly because it maintains many Soviet-era policies (e.g., no privatization); Uzbekistan is a major cotton exporter, but environmental degradation threatens cotton production; it

has significant gold and natural gas deposits and chemical and machinery factories from Soviet era

4. Kyrgyzstan moved aggressively to privatize state-run industries, but its economy is largely agricultural and few of its industries are competitive; since independence, its economy has shrunk by 50 percent; Kyrgyzstan has the largest supply of fresh water in the region

5. Turkmenistan has a substantial agricultural base because of Soviet irrigation projects, and exports cotton; it continues its state-run economy; Turkmenistan hopes development of oil and natural gas fields will bring prosperity

6. Azerbaijan has the region's best-developed fossil fuel industry; it has attracted much international interest and investment, but its economy has been slow to respond, and Azerbaijan remains a very poor country

7. Tajikistan is the most economically troubled of the former Soviet republics, with per capita GNI of only $280; Tajikistan is remote, with poor transportation links to the outside, most of them to Russia; greater efficiency is needed; civil war has made matters worse

8. Mongolia was a Soviet ally and was run by a communist party; Mongolia no longer receives Russian foreign aid; its agricultural foundation is meager; and its traditional trade in livestock is not very profitable; isolation is another problem

C. The Economy of Western China
   1. Chinese portions of this region have not suffered as much as former Soviet areas; although China's population is growing rather quickly overall, most growth is on the east coast
   2. Tibet is one of the world's poorest places, partly because of its isolation; but many Tibetans are able to cover most of their basic needs; immigration of Han Chinese to the region threaten to upset local balance between human numbers and the environment in western China
   3. Xinjiang has tremendous mineral wealth (Fig. 10.27); agriculture is productive; indigenous Muslims resent influx of Han Chinese

D. Economic Misery in Afghanistan: Afghanistan is the poorest in the region, and has been at war since the late 1970s (more than 20 years); it was poor before the war, and war has made poverty far worse; few legitimate exports (animal products, hand-woven carpets, fruits, nuts, semiprecious gems); most energy and consumer goods are imported; Afghanistan became a major exporter of opium (heroin) during the 1990s; United States bombs have caused more damage, but the United States and other countries are pledging billions in economic aid to Afghanistan

E. Central Asian Economies in Global Context
   1. Afghanistan firmly embedded in world economy (weapons and drugs)
   2. Former Soviet republics maintain relations with Russia, and are establishing relationships with

other partners (e.g., Iran, Pakistan, Turkey)
3. United States and other Western countries interested in the region's oil and natural gas
4. Fossil fuel reserves create a complex international economic environment, including a need for new pipelines, but each prospective route has drawbacks; in 2001, Azerbaijan, Kazakstan, Georgia, and Turkey signed a memorandum of understanding supporting a U.S.-proposed route
F. Social Development in Central Asia
1. Social Conditions and the Status of Women in Afghanistan: Afghanistan has suffered through warfare since 1978; it has high infant and childhood mortality; average life expectancy is only 45 years; only about 15 percent of Afghan women are literate; under Taliban control, women were prohibited from working outside the home, attending school, and often from obtaining medical care; today Taliban forces have attacked several girls' schools; nearly one-third of Afghans have been refugees over the past decade; now refugees are beginning to return
2. Social Conditions in the Former Soviet Republics and Mongolia: women in these republics have a higher social position than Afghan women, largely because of Soviet policies; compared to other parts of the region, the socioeconomic indicators are the most favorable in the region; eastern China is improving, but Chinese territories in Central Asia (the western part of the country) may not benefit; 60 percent of non-Han Chinese are illiterate; although some ethnic peoples in this region may be exempt from Chinese one-child policies, reports of forced sterilizations and abortions create tensions in the area; with economic collapse in the region, these statistics may not hold at their current levels; in Mongolia, women are well-represented in professions, but have little political power
3. Social Conditions in Western China: current, accurate statistics on China are hard to find

VII. Conclusions

Dominated for many years by Russia and China, this region is only now emerging as a separate entity. It is a region with rugged terrain and a historically pastoral society. Today the presence of fossil fuels is generating worldwide interest in the region. This is a region that is experiencing difficult times, owing largely to the collapse of political and economic systems at the beginning of the 1990s. As well, warfare and other armed conflicts have damaged the economy of the states of this region, especially Afghanistan, which has been embroiled in war since the 1980s. Afghanistan has emerged as a focus of United States and world interest since September 2001.

PRACTICE MULTIPLE CHOICE QUIZ

1.   What effect has Central Asia's low population density had on the region's environment?
     a.   In spite of the low population density, the fragility of the environment has caused it to become highly polluted
     b.   Low population density has had no effect on the environment
     c.   Low population density has helped keep the environment relatively clean
     d.   Low population density has made it difficult for the countries of Central Asia to persuade Russia to take actions to reduce acid rain coming from Russia into the region
     e.   The many old, polluting factories in the region have negated the effects of low population density on the region, leading to a devastate environment

2.   Why was the Aral Sea particularly sensitive to irrigation development, resulting in an enormous loss of the lake's surface area?
     a.   It began as a small lake (by volume)
     b.   It contains freshwater, making it a direct source of water for many users
     c.   It is located in the center of a major population center, with several major cities on its banks
     d.   It is relatively shallow and its tributaries flow across a vast agricultural landscape
     e.   All of the above

3.   All of the following statements about the physical geography of Central Asia are correct, EXCEPT...
     a.   Central Asia includes a central belt of deserts
     b.   Most of Central Asia is arid
     c.   Most of Central Asia is characterized by plains and basins of low and intermediate elevation
     d.   The highlands of Central Asia originated as the result of volcanic eruptions
     e.   The mountains of Central Asia are higher and more extensive than those found anywhere else in the world

4.   How do most people of Central Asia make a living?
     a.   As computer software developers
     b.   As laborers in the region's factories
     c.   As pastoralists
     d.   As traditional hunters and gatherers
     e.   As workers on cotton plantations

5.   Conquest of Central Asia by which neighboring countries fostered the establishment of cities in the region?
     a.   China and Russia
     b.   India and Bangladesh
     c.   Iran and Pakistan
     d.   Nepal and Bhutan
     e.   Turkey and Georgia

6. What is the dominant linguistic group in Central Asia?
   a. Chinese
   b. Indo-European
   c. Mongolian
   d. Tibetan
   e. Turkic

7. Which country of Central Asia is essentially a theocracy?
   a. Kazakhstan
   b. Mongolia
   c. Tibet
   d. Uzbekistan
   e. All of the above

8. Which foreign language was common in Central Asia during the Soviet era?
   a. Cantonese   b. English   c. German   d. Mandarin Chinese   e. Russian

9. Which two countries of Central Asia have not been under direct control by either the Russians or the Chinese?
   a. Armenia and Kyrgyzstan
   b. Kazakhstan and Azerbaijan
   c. Mongolia and Afghanistan
   d. Tibet and Uzbekistan
   e. Turkmenistan and Tajikistan

10. What was the primary means by which early Soviet leaders (like Vladimir Lenin) tried to protect non-Russian peoples in Central Asia from Russian domination?
    a. Creation of "union republics" with some autonomy in Central Asia
    b. Enactment of laws promoting local languages and voting rights
    c. Establishment of local militias in the Central Asia republics to repel the Russian military
    d. Passage of laws forbidding encroachment by Russians in to Central Asia
    e. All of the above

11. At what point did China reclaim most of its Central Asian territories?
    a. After China re-emerged as a united country in 1949
    b. After the breakup of the Soviet Union around 1990
    c. At the end of the Vietnam War
    d. At the end of World War I
    e. During the Q'ing Dynasty

12. Beginning in 1978, which country of Central Asia has experienced war with Russia, followed by internal conflict, then United States occupation, which continues today.
    a. Afghanistan   b. Azerbaijan   c. Mongolia   d. Tibet   e. Uzbekistan

13. Based on conventional measures, how would one rank Central Asia in terms of its economic development?
    a. In the middle income group of countries
    b. Moderate economic development
    c. One of the least prosperous in the world
    d. One of the most prosperous regions in the world
    e. Data are not available to make such assessments

14. What is the most prosperous country in Central Asia?
    a. Afghanistan   b. Kazakhstan   c. Mongolia   d. Tajikistan   e. Uzbekistan

15. What economic activity will most likely link Central Asia to the global economy in the future?
    a. Agricultural products
    b. Assembly plants
    c. Foreign direct investment
    d. Oil and natural gas
    e. Software programming

Answers: 1-C; 2-D; 3-D; 4-C; 5-A; 6-E; 7-C; 8-E; 9-C; 10-A; 11-A; 12-A; 13-C; 14-C; 15-D

Chapter 11
EAST ASIA

LEARNING OBJECTIVES
- This chapter describes East Asia, which includes China, Japan, South Korea, North Korea, and Taiwan.
- Students should understand the historical cultural cohesiveness of East Asia, while recognizing the evolution of economic and political diversity at the end of the 20th century in this region.
- Students should also understand the world prominence of the countries in this region in a historical sense and understand the causes of the periodic rise and decline of the countries of this region. As well, students need to understand the current role of this region in world trading patterns.
- Upon completion of this chapter, students should be familiar with the physical, demographic, cultural, political, and economic characteristics of East Asia.
- In addition, the student should understand the following concepts and models:

  - Loess soils
  - Confucianism
  - Central place theory
  - Ideographic writing

CHAPTER OUTLINE
I. Introduction
East Asia is the world's most populous region, with over 1.5 billion people and six countries. This region exhibits a historical pattern of cultural unity and periods of isolationism. China is both the oldest culture in East Asia as well as the oldest continuous national culture in the world. Japan is a much newer state, but since its inception, the rivalry between it and China has been a recurring regional theme. This region has experienced colonization, and in the 20th century has seen both internal and international conflict. Economically this region includes one of the world's wealthiest countries (Japan) but also includes extreme poverty.

II. Environmental Geography: Resource Pressures in a Crowded Land
A. General Description (Fig. 11.2)
1. East Asia's environmental problems are unusually severe because of: (1) large population, (2) rapid industrialization, (3) unique physical geography
2. The wealthier countries (such as Japan) have invested in environmental protection; other countries (including China) do not have the resources for environmental protection
B. Dams, Flooding, and Soil Erosion: China's Yangtze (or Chang Jiang) River is an important feature, and is now subject to environmental controversy

1. The Three Gorges Controversy: the Chinese government wants to control the Yangtze for two reasons, to (1) prevent flooding and (2) generate electricity; Three Gorges Dam is now under construction to achieve these goals; this will be the world's largest hydroelectric dam, and will displace between 1.2–1.9 million people (largest forced resettlement for a single project in history) and cost $25 billion; the dam will threaten several endangered species (including the Yangtze River dolphin); the project is so environmentally destructive, World Bank withdrew support; private funding (from China and the United States) will fill the gap

2. Flooding in Northern China: a deforested region, it has suffered from drought and flood; the Huang He has the worst floods and carries a huge *sediment load* (the amount of suspended clay, silt and sand in water); it is the world's muddiest river; the Huang He's deposition of soils during flood cycles has created the North China Plain

3. Erosion on the Loess Plateau; located to the west of the North China Plain, erosion from the Loess Plateau is the source of Huang He's sediment deposits; erosion has reduced farmland on the Loess Plateau; efforts to stop the erosion have been ineffective, and many people are migrating from the region

C. Other East Asian Environmental Problems

    1. Forests and Deforestation: most of uplands of China and South Korea support only grass, meager shrubs, and stunted trees; China's forestry practices have not been environmentally sound; China has a severe shortage of forest resources and may need to import lumber, pulp, and paper in the near future

2. Mounting Pollution: as China's industrial base expands, water pollution and toxic-waste dumping are becoming more serious problems; Taiwan and South Korea experienced these problems earlier, and are introducing stricter environmental regulations, and (like Japan) setting up factories in poor places with less-restrictive environmental standards

3. Environmental Issues in Japan: Japan is densely settled; this has some benefits, including more efficient public transit; in the past, Japan suffered pollution, and several toxic-waste disasters that killed thousands of people; these disasters led to strict air and water pollution laws; Japan (like other countries) practices *pollution exporting* (relocating dirtier factories to other countries)

4. Endangered Species: East Asia is a major center for trade in endangered species; many traditional Chinese medicines are derived from rare or exotic animals (e.g., deer antlers, bear gall bladder, tiger penises,

rhinoceros horns, snake blood); China has little remaining wildlife, but is trying to protect its remaining habitat, including the bamboo thickets that pandas need

D. Global Warming and East Asia
1. China probably surpassed the United States in greenhouse gas emissions in 2007 because of economic growth and the use of coal, and has begun to address environmental issues
2. Rising temperatures could cause a decrease in production of corn, wheat, rice
3. Economic development takes precedence over the environment in China
4. Japan, South Korea, Taiwan are big contributors to greenhouse gases, but they are energy efficient

E. East Asia's Physical Geography (Fig. 11.7)
1. East Asia is in the same latitudinal range as the United States; lies at the intersection of three tectonic plates and is geologically active
2. Japan's Physical Environment: Japan is comprised of four major islands; its climate is temperate, although the south is subtropical, and the north is almost subarctic; the Pacific coast of Japan is subject to *typhoons* (hurricanes); Japan has a rugged terrain, with mountains over some 85 percent of its land area; because of a favorable climate and a long history of forest conservation, Japan is the world's second most heavily forested industrialized country

3. Taiwan's Environment: about the size of Maryland; the central and eastern regions are rugged and mountainous; the west is dominated by alluvial plains; Taiwan straddles the Tropic of Cancer and has a mild winter climate; it is subject to typhoons in early autumn
4. Chinese Environments (Fig. 11.12): Southern China has rugged mountains and hills interspersed with lowland basins; south of the Yangtze Valley, the mountains are higher but still interspersed with alluvial lowlands; north of the Yangtze Valley, the climate is colder and drier than to the south; summer rainfall is plentiful, except at the edge of the Gobi desert; the North China Plain is cold and dry in winter and hot and humid in summer; parts of this area are threatened by desertification; the Loess Plateau is a fairly rough upland of moderate elevation and uncertain precipitation; Manchuria is dominated by broad lowlands sandwiched between mountains and has harsh winters and warm, moist summers
5. Korean Landscapes: Korea is a peninsula, partly cut off from Manchuria by rugged mountains and sizable rivers; Korea itself is mountainous with scattered alluvial basins; the north has abundant mineral deposits and forest and hydroelectric resources; large portions of South Korea suffered deforestation

III. Population and Settlement: A Realm of Crowded Lowland Basins (Fig. 11.15)

A. East Asia and South Asia are the most densely populated regions on Earth
   1. The region's population growth has declined dramatically since the 1970s
   2. Japan is facing population loss, with an aging population and fewer young people (low birth rate)

B. Agriculture and Settlement in Japan (127 million people)
   1. Japan is highly urbanized; agriculture shares limited lowlands with cities and suburbs
   2. Japan's Agricultural Lands: mostly in coastal plains and interior basins; Japan's most important crops are rice, vegetables, and fruit
   3. Settlement Patterns: all Japanese cities are located in the same lowlands (15% of Japan's total land area) that support Japan's farms; Tokyo, Osaka, Nagoya are all at the center of three largest plains; density is high—338 people/sq km (compared to United States at 31/sq km)
   4. Japan's Urban-Agricultural Dilemma: While cities are settled at very high densities, allowing them to sprawl into adjacent agricultural land would eliminate farms needed to help feed Japan's people; Japan protects its farmers to ensure food self-sufficiency to the extent possible, even though it means higher rice prices for Japanese consumers

C. Settlement and Agricultural Patterns in China, Taiwan, and Korea
   1. Taiwan, Korea are urban; China is mostly rural (70 percent); what cities exist are fairly evenly distributed in China's plains and valley
   2. China's Agricultural Regions: southern and central China's population is centralized in lowlands, where summer rice and winter barley alternate, and there are sometimes three crops per year; North China Plain is one of the most *anthropogenic landscapes* (one that has been heavily transformed by humans); Manchuria is thoroughly settled, but is less crowded than other parts of China and produces a surplus of food; the Loess Plateau is arid, exhibits erosion, and is thinly settled by people who grow wheat and millet at subsistence levels; housing in the Loess Plateau includes underground houses that are cool in summer, warm in winter, but may collapse during earthquakes
   3. Patterns in Korea and Taiwan: Korea is densely settled at 487 people per square km, more than Japan (23 million in North Korea where corn dominates, 486 million in South Korea, where rice prevails); Taiwan is even more densely settled (636/square km, total population of 23 million) with most living in lowlands on the north and west coasts in an area that contains large cities, factories, and lush farms

D. Agriculture and Resources in Global Context
   1. Agriculture in the region is productive, but does not produce

116

enough to feed its large, densely settled population

2. Global Dimensions of Japanese Agriculture and Forestry: Japan is one of the world's largest food importers (meat, wheat for bread and noodles, and feed for livestock from the United States, Canada, Australia; soybeans from the United States, Brazil; farm-raised seafood from Southeast Asia and Latin America); Japan also imports lumber and pulp (United States, Southeast Asia, Latin America), energy, and minerals; because Japan's industry produces valuable items for export, it can afford to purchase these imports

3. Global Dimensions of Chinese Agriculture: China was self-sufficient in food production through the 1980s, but now must import food; China's government promises to boost production soon, but by the year 2030, China may have to import 5–10 percent of its total grain demand

4. Korean Agriculture in a Global Context: like Japan and Taiwan, South Korea has a global resource procurement pattern, importing large quantities of wheat and corn; North Korea has a strict policy of food self-sufficiency, but in years of low productivity (such as 1997–98), famine is the result

E. Urbanization in East Asia
1. East Asia had a well-developed system of cities early on, but remained largely rural at the end of WWII
2. Chinese Cities: China's traditional cities were designed according to strict geometrical principles, with a uniform structure, surrounded by defensive walls and horizontal, low buildings; when Europeans colonized China, they took existing port cities and constructed Western-style buildings and modern business districts; when communists came to power in 1948, colonial city Shanghai was the world's second-largest city and an example of Western decadence to be milked for taxes that could be invested elsewhere; today Shanghai remains an important Chinese city, second to Beijing; Beijing, China's capital, was reshaped when communists came to power; some important historical districts survived (e.g., Forbidden City).

3. The Chinese Urban System: because of China's indigenous heritage of urbanism and socialist planning under the communists, China's urban system is balanced, with large cities evenly spaced across the inhabited areas, and no city overshadowing all others, as described in Walter Christaller's *Central Place Theory* (the principle that an evenly distributed rural population will give rise to a regular hierarchy of urban places, with uniformly spaced larger cities surrounded by constellations of smaller cities, each of which is surrounded by smaller towns, all because of retail marketing and the need for people to have ready access to a variety of goods and services)

4. Other Regional Urban Patterns: *urban primacy* is evident in

South Korea (Seoul); Japan exhibits *superconurbation:* megalopolis, a huge coalesced metropolitan area; Japanese cities are mostly new, since most pre-modern and historical buildings were made of wood and were destroyed by bombs or fire in WWII

IV. Cultural Coherence and Diversity: A Confucian Realm?

A. East Asian commonalties go back to ancient Chinese civilization, about 4,000 years ago, remaining relatively isolated and self-contained until the 1800s

B. Unifying Cultural Characteristics

1. Religious and philosophical beliefs of Buddhism and Confucianism have shaped both individual beliefs and social and political structures

2. The Chinese Writing System: based on *ideographic writing,* where each symbol (called an ideograph or character) represents primarily an idea rather than a sound; there are thousands of characters representing ideas used in the language; written and spoken language are largely separate; the major disadvantage of ideographic writing is that the large number of symbols makes it difficult to learn; the major advantage is that two literate people who understand the ideographs can read any language that shares the same symbols; Chinese ideographs spread to Japan (with major modifications), and Korea and Vietnam (which later switched to an alphabet system)

3. Korean Modifications: in the 1400s, Korea adopted an alphabet system to improve literacy and differentiate Korean culture from Chinese

4. Japanese Modifications: major grammatical differences between Japanese and Chinese made use of symbols (or *kanji,* as Japan calls them) awkward; the Japanese developed *hiragana* (a unique quasi-alphabet, called a syllabary, in which each symbol represents a distinct syllable or combination of a consonant and vowel sound); because of lingering similarities between written Chinese and Japanese, many literate Japanese people can read Chinese; Japanese can also be written using the Roman alphabet (the alphabet you're reading right now, also called *romanji* in Japanese)

5. The Confucian Legacy: Confucianism refers to the philosophy (NOT a religion) of Confucius; it is important throughout East Asia, but more so in China and Korea; in the sixth century B.C., Confucius stressed (1) deference to legitimate authority; (2) need for authority to be responsible and benevolent (emperor could be replaced if he failed to fulfill duties); (3) a patriarchal family system where children obey and respect elders (especially older males); (4) need for a well-rounded and broadly humanistic education; (5) value of meritocracy in which people should be judged on their behavior and education, not their family connections, which led to

selection of high officials of imperial China *(Mandarins)* based on competitive exams; of course, usually only the wealthy could afford the education that would lead to success on exams; Confucianism held that the emperor was an almost godlike father figure for the entire country

6. Confucianism in Japan: never as important as on the mainland; Japanese excluded beliefs they considered dangerous (e.g., the clause that allowed the Chinese to replace a leader who did not fulfill his responsibility); as a result, Japan has had the same ruling family throughout its written history

7. The Modern Role of Confucian Ideology: old conventional wisdom—Confucianism's conservatism and respect for tradition and authority—caused East Asia to be economically backward; new conventional wisdom: Confucianism's respect for education and social stability gave East Asia a tremendous advantage in international competition and helps explain the region's strong economic growth after WWII

C. Religious Unity and Diversity in East Asia

1. Mahayana Buddhism: provides East Asia's most important culturally unifying beliefs; stresses the human soul's quest to escape an endless cycle of rebirths and reach *nirvana* (or total enlightenment and union with the divine cosmic principle); founded in India around 600 B.C., now widespread in East Asia;

Mahayana Buddhism is nonexclusive, allowing people to follow it while professing belief in another faith; variations include Japanese Zen, which demands that followers engage in "mind emptying"; at various times, Buddhists have been persecuted in East Asia because the religion originated outside the region

2. Shinto (Japan): began as an animistic worship of nature spirits, but now is a subtle set of beliefs about harmony of nature and humanity; Shinto is a place- and nature-centered religion that considers certain places sacred (e.g., Mt. Fuji)

3. Taoism and Other Chinese Belief Systems: stresses the importance of spiritual harmony and the pursuit of a balanced life, it is indirectly associated with *feng shui* (or *geomancy*), which is the Chinese and Korean practice of designing buildings in accordance with spiritual power that some believe courses through the local topography; other traditional religions focus on unique attributes of particular places *(particularlism),* but these religions are usually practiced only locally

4. Minority Religions: Protestant Christianity is growing in Korea and possibly in China; China has large populations of Muslims (called *Hui*) in northwest China, in Yunan province (south-central border), and in scattered villages in nearly every province of China

5. Secularism in East Asia: East Asia is one of the most secular regions on earth; in Japan,

religion is not very important; Confucianism is a philosophy, NOT a religion; Marxist philosophy in China suppressed religious practice there; but as enforcement has eased, religious expression is returning; this is not the case in North Korea, which still rigidly enforces orthodox Marxism (communism)

D. Linguistic and Ethnic Diversity (Fig. 11.25)
1. While written, ideograph-based language may unify the region, spoken language is very different
2. Language and National Identity in Japan: Japanese is the only member of its language group; Japanese are mostly homogeneous and generally view themselves as a unified people; at one time, there were two distinct groups (Japanese in the south and Ainu in the north), and the Japanese drove out most Ainu
3. Minority Groups in Japan: there are small differences in Japanese dialects; about 700,000 ethnic Koreans may face discrimination in Japan and are rarely able to obtain Japanese citizenship; since the 1980s, immigrants have come to Japan, most from poorer Asian countries, but few become permanent residents, let alone citizens; Burakamin, Japan's indigenous outcast group whose ancestors worked in "polluting" industries (like leathercraft), once faced discrimination and remain among the poorest and least educated people in Japan
4. Language and Identity in Korea: Koreans are largely homogeneous, but there is a strong sense of regional identity that traces its roots to medieval kingdoms of the country; many Koreans live in China, Kazakhstan, and continue to migrate to the United States, Canada, Australia, and New Zealand
5. Language and Ethnicity Among the Han Chinese: Han Chinese form the vast majority of people who have long been incorporated into Chinese cultural and political systems and whose language is expressed in the Chinese writing system, though they do not all speak the same language; the Hakka, who speak a southern Chinese language, are sometimes considered not to be a true Han Chinese; Hakka fled northern China 1,000 years ago to southern China, where they grow crops and work as loggers, stonecutters, or metal workers; many Hakka consider themselves Han; Chinese language is *tonal* (word meaning changes according to the pitch in which a person says it)
6. The Non-Han peoples: tribal groups live in more remote upland areas of China; tribal implies a group of people with a traditional social order based on self-governing village community; many of them once had their own kingdoms but are now subject to Chinese state; the land area of these people has diminished over time; in Guangxi, most people speak languages in the Tai-Kadai family (Guangxi is an autonomous region, designed to allow non-Han peoples to experience "socialist

modernization" at a different pace from the rest of China)

7. Language and Ethnicity in Taiwan: tribal peoples speak languages related to those of Indonesia; Han Chinese migrated to Taiwan around 16th century and their language evolved into Taiwanese; nationalist Chinese speaking Mandarin migrated to Taiwan in 1949; tension resulted between the two language groups

E. East Asian Cultures in Global Context

1. Tension exists between internal orientation and tendencies toward cosmopolitanism

2. The Cosmopolitan Fringe: found in large cities, includes Western influences such as English language studies and college study abroad, Internet usage; the contributions of this fringe to global culture include Hong Kong action films, Japanese dominance in electronics, videogames, and the auto industry; Japanese ultra-nationalists call for their fellow citizens to resist Western decadence and retain traditions of the *samurai* (the warrior class of pre-modern Japan), while others worry that Japan is too insular

3. The Chinese Heartland: China has historically been self-sufficient and insular; major influences have been traditional Chinese society and communism in the last half of the 1900s; opening of China in 1970s and 1980s has increased the prominence of the southern coast and allowed the reemergence of regional and local identities

V. **Geopolitical Framework: Enduring Cold War Tensions** (Fig. 11.31)

A. A key feature of East Asia's political history is the centrality of China and Japan's ability to remain outside China's reach

1. China's view had two categories: places were either part of the Chinese empire, or totally outside it, as barbarians

2. There is a long-term rivalry between China and Japan

B. The Evolution of China

1. Chinese culture hearth (1800 B.C.) was North China Plain and the Loess Plateau; China was ruled by a series of dynasties (families) from 219 B.C. until 1912, controlling roughly the same territory; China ruled regions to its west, but never incorporated them into China; China tried to conquer Korea, but Korea retained its freedom by paying tribute and acknowledging Chinese supremacy; for most of the past 2000 years, China was the wealthiest and most powerful state on the planet; the only real threat came from Mongolia and Manchuria; often conquering armies adopted Chinese customs in order to govern the far more numerous Han people

2. The Manchu Qing Dynasty: the last of China's dynasties, ruling from 1644 to 1912; it functioned well until the middle 1800s; then European (especially British), later Japanese forces began to dominate; under the Manchu Dynasty, China had its largest territorial extent and was at its most powerful

3. The Modern Era: Chinese Empire was in decline in the 19th century, as it failed to keep pace with technological and economic progress of Europe; initially, Chinese leaders did not perceive a threat from European merchants on the Chinese coast; when Europeans found the price of Chinese silks too high and saw that Chinese did not desire the products they had for sale, the British began to sell opium in China; China wanted to eliminate opium and battled England in two "opium wars" to achieve this objective, but lost; as British dominance increased, political unrest occurred and eventually toppled the Manchu Dynasty in 1911; from then until after WWII, nationalists and communists competed for control of China; at the end of WWII, communists gained control, pushing the nationalists to Taiwan

C. The Rise of Japan
1. Japan emerged as a unified state more than 2,000 years later than China, during the seventh century; between A.D. 1000 and A.D. 1580, Japan was separated into many mutually antagonistic feudal realms
2. The Closing and Opening of Japan: in early 1600s, Japan reunited by armies of the Tokugawa Shogunate (a *shogun* is a military leader who theoretically remains under the emperor, but who really holds political power); until the 1850s, Japan traded only with China, Russia, and the Dutch; in 1853, United States gunboats in Tokyo Bay demanded access to Japan; Japan repelled them, but then modernized its economic, administrative, and military systems to protect itself in the future; this modernization was called the *Meiji Restoration,* and was ultimately very successful

3. The Japanese Empire: fueled by the Meiji Restoration, Japan developed a silk industry for export in order to raise money to purchase modern equipment; Japan also set on a course of expansionism to meet the challenge of its day; Japan took control of Hokkaido, Kuril, and Sakhalin islands, Taiwan; gained influence in northern China, and annexed Korea in 1910; Japan was allied to Great Britain, France, and the United States in WWI, and won Germany's Micronesian colonies; in the 1930s, Japan occupied Manchuria, the North China Plain and coastal cities of Southern China; in 1941, Japan decided to destroy the United States naval fleet in the Pacific Ocean in order to clear the way for conquest of resource-rich Southeast Asia, which would unite East and Southeast Asia in a "Greater East Asia Co-Prosperity Sphere," ruled by Japan and keep out United States and Europe; Japanese were sometimes brutal in their treatment of Chinese and Koreans; ultimately, Japan was stopped (WWII)

D. Postwar Geopolitics
1. With its defeat in WWII, Japan lost its colonial empire, but Japan has bounced back

2. Japan's Revival: after WWII, Japan retained control of its own four islands and some other minor islands; after losing its possessions, Japan was forced to rely on trade to rebuild its economy and infrastructure; although Japan's post-war constitution required that it use the United States military rather than one of its own, slowly it has been rebuilding its military

3. The Division of Korea: U.S.S.R and the United States divided Korea after WWII; land north of the 38th parallel went to U.S.S.R., land south of the 38th parallel went to the United States, creating two separate regimes; North Korea invaded South Korea in 1950; United States aided the South, China helped the North; the war ended in stalemate, still divided, but there are efforts to reconcile; North Korea may not be living up to its commitments, so reunification is questionable; South Korea's economy has grown and developed

4. The Division of China: conflict that arose in China after the decline of the Manchu Dynasty in 1911 was resolved in 1949 when the communists won and forced the nationalists to flee to Taiwan; mainland China is now called People's Republic of China; Beijing still claims China, and vows to reclaim it, while Taiwan maintains that it is the true government of China; the United States recognized Taiwan as the legitimate government in the 1950s and 1960s, then in the 1970s, recognized People's Republic of China; many Taiwanese would like to declare independence, but if it does so, China has threatened to invade it and take it back by force; in 2001 both China and Taiwan were admitted to the World Trade Organization

5. Chinese Territorial Issues: today China controls Tibet, Xinjiang (which local people call Eastern Turkestan), and Inner Mongolia; China also has territorial conflicts with Russia and India; also, Hong Kong, which was controlled by the British, reverted to Chinese control in 1997, and Macau (a Portuguese colony) reverted to Chinese control in 1999

E. Global Dimensions of East Asian Geopolitics

1. East Asia was caught up in the Cold War as China and North Korea were seen as friends of the Soviet Union, while Japan, Taiwan, and South Korea were friends of the United States; this is no longer the case

2. The North Korean Crisis: North Korea has acknowledged that it has nuclear weapons. Talks involving North Korea, South Korea, Japan, China, Russia, and the United States have taken place, but there is no resolution yet.

3. China on the Global Stage: China has become a strong military power, and many United States firms have factories there. United States has criticized China for its human rights violations; China says United States has many such violations.

4. A clash of civilizations? Will China and East Asia become a world-leader power bloc? Certainly it is becoming a more central actor.

VI. Economic and Social Development: An Emerging Core of the Global Economy

A. Economic and Social Development in East Varies from Country to Country

B. Japan's Economy and Society

1. Japan was the world's economic pacesetter in the 1960s, 1970s, and 1980s but has seen decline in the 1990s; even so, it is still the world's second largest economic power

2. Japan's Boom and Bust: the central government of Japan plays a much larger role in the country's economy than occurs in the United States; groups of companies in the same or related industries (keiretsu) are intertwined and would be considered as cartels (which are illegal) if they were in the United States; advantages of the Japanese system: supports stability, promotes long-term planning, and minimizes layoffs; disadvantages: reduces flexibility, causes higher consumer prices and low corporate profits; has done nothing to improve Japanese agricultural efficiency

3. The Japanese Economic System: Close cooperation between government and business, and among businesses to promote stability and long term stability

4. Living Standards and Social Conditions in Japan: Japan has higher GNP than the United States in 1999, but living standards are slightly lower than in the United States because of higher prices for housing and food; unemployment in Japan is lower, and there is no extreme poverty, but professional and managerial jobs are noted for their long hours and high stress; crime rates are low; Japan's literacy, infant mortality, and average life expectancy are better than those in the United States

4. Women in Japanese Society: career opportunities for women are limited; women are delaying marriage; and low fertility rates threaten to cause population decline in Japan

C. The Newly Industrialized Countries (South Korea, Taiwan, Hong Kong)

1. The Rise of South Korea: in the 1960s, the government established a program of export-led economic growth that led to a transition from inexpensive consumer goods to heavy industrial products to high-tech equipment; South Korea has invested in its educational system

2. Contemporary South Korea: in the 1960s and 1970s, South Korea's government was repressive; freedom came in the 1980s; in the 1990s, scandal and corruption in government and business caused political turmoil; economic difficulties plague South Korea, and government and economic reforms may be needed

3. Taiwan and Hong Kong: Taiwan's economy is organized around small and mid-sized family farms, which makes the

124

economy more flexible and agile, while making it difficult to enter industries that require much capital; Taiwan was not strongly affected by the "Asian crisis" in the economy of the late 1990s; Hong Kong has a *laissez-faire* economic system (complete market freedom without government control); Hong Kong specializes in business services, banking, telecommunications, and entertainment

D. Chinese Development

1. China under Communism: China nationalized its industry after 1949; the Great Leap Forward of the late 1950s used small-scale village workshops to produce large quantities of iron, causing peasants to melt down their farm tools to meet unreasonable quotas and contribute large proportions of their crops to the state, resulting in famine, with as many as 20 million people starving to death; in the 1960s, the Cultural Revolution mobilized young people to eliminate any remaining vestiges of capitalism; many well-educated people were sent to rural villages for "reeducation" through hard physical labor, while others were killed, devastating the economy

2. Toward a Post-Communist economy: leader Mao Zedong died in 1976, ushering in more pragmatic leaders who have fostered somewhat closer Chinese ties with the world economy; while no longer exclusively communist (socialist), China is not truly capitalist either; China still

operates most heavy industries, and controls the economy

3. Industrial Reform: China created *Special Economic Zones (SEZ)* (areas where foreign investments are welcome and state interference minimal); China's economy is growing at 8–15 percent per year; it generates large trade surpluses, especially with the United States; one of the world's major trading nations in 1990s; China joined the WTO in 2001

4. Social and Regional Differentiation: China's reforms have resulted in *social and regional differentiation* (occurs when certain groups of people, and certain parts of the country, prosper while others falter)

5. The Booming Coastal Region: the greatest economic growth in China occurs in coastal cities and Beijing (Guangdong and Fujian)

6. Interior and Northern China: Manchuria is a *rust belt* (a zone of decaying factories), but remains relatively prosperous because of its fertile soils; many interior provinces are experiencing economic decline and outmigration; rapid and uneven economic growth has caused problems (e.g., inflation, government corruption, rising crime, and organized crime)

7. Rising Tensions: Rapid and uneven economic growth has caused problems (e.g., inflation, government corruption, rising crime, and organized crime); China's people struggle for democratic reform, individual freedoms; Chinese government has outlawed the Falun Gong, a

quasi-Buddhist group devoted to meditation and exercise, claiming that it was an evil cult that threatened China; practitioners of Falun Gong risk imprisonment or worse. Other countries complain about China's human rights violations, trade surplus, and have accused it of pirating music, software, and brand names.

E. Social Conditions in China: Communists made progress in several areas (health and education), but regional disparities continue

　1. China's Population Quandary: China has 1.3 billion people, concentrated in a few areas; in the 1980s, China instituted a one child policy, which was successful in reducing total fertility rate to 1.6; the policy, coupled with a strong preference for male children, has had negative unintended consequences; among the extreme measures some people have taken to ensure a baby boy are (1) gender-selective abortion, (2) kidnapping a baby boy, (3) abandonment of baby girls, (4) female infanticide, all resulting in an imbalance between the sexes (a higher than expected number of boys)

　2. The Position of Women: women have traditionally had low status in China; communists and nationalists alike worked to improve the status of women; female participation in the labor force has increased, but few women have power either in business or in government

VII. Conclusions

East Asia has close cultural and historical bonds. China, the world's oldest continuous culture, has played a key role in East Asia, and its rivalry with Japan has been especially relevant in the recent past. The desire of many non-Han peoples in China for independence, economic disparities among the regions of China, along with a growing demand for democracy and freedom of expression, present a challenge to the ruling Communists over the coming years. Japan's recent economic difficulties, along with those of Taiwan and South Korea, also suggest an uncertainty in the coming years. Still, it is likely that this region will continue to play a key role in the economy and politics of the global system for the foreseeable future.

PRACTICE MULTIPLE CHOICE QUIZ

1. Which of the following statements about China's Three Gorges Dam is TRUE?
    a. The Three Gorges Dam will be built on the Yangtze River
    b. It will displace more than one million people
    c. It will help prevent flooding and be used to generate hydroelectricity
    d. The ecological and human rights consequences of the Dam are so negative that the World Bank withdrew its support
    e. All of the above

2.	Which country of East Asia has a very long history of forest conservation?
	a. China	b. Japan	c. North Korea	d. South Korea	e. Taiwan

3.	With what other land use does Japan's agriculture compete for space in the country's coastal plains and interior basins?
	a. Cities and suburbs
	b. Government buildings
	c. Industrial areas
	d. Open space
	e. Retail activities

4.	How has the diet of Japanese people been able to grow more diverse?
	a. Development of new hybrid crops
	b. Genetically engineered crops
	c. Importation of food
	d. Joint agricultural ventures with the United States
	e. All of the above

5.	Only one country of East Asia is still mostly rural; which one is it?
	a. China	b. Japan	c. North Korea	d. South Korea	e. Taiwan

6.	Which geographic theory best explains the distribution of Chinese cities?
	a. Central Place
	b. Concentric Zone
	c. Counterurbanization
	d. Historical Evolution of the City
	e. Urban Realms

7.	What is the basis for the Chinese ideographic writing system?
	a. Syllables	b. Phonetic sounds	c. Ideas, symbols	d. Hieroglyphics	e. Greek

8.	What religion is most widespread in East Asia?
	a. Animism	b. Buddhism	c. Christianity	d. Hinduism	e. Islam

9.	Up until the mid-1980s, what was the approach of the countries of East Asia toward the West?
	a. They adopted Western technology
	b. They attempted to insulate themselves from Western cultural values
	c. They cautiously worked with Western countries
	d. They embraced Western culture
	e. They traveled to the West in search for education

10.	The political history of East Asia revolves around the centrality of which country in the region?
	a. China	b. Japan	c. North Korea	d. South Korea	e. Taiwan

11. What country of East Asia has been a source of concern because of its development of missiles that are capable of carrying nuclear warheads?
   a. China   b. Japan   c. North Korea   d. South Korea   e. Taiwan

12. Which other country of East Asia does the government of China consider to be an integral part of China?
   a. Hong Kong   b. Japan   c. North Korea   d. South Korea   e. Taiwan

13. What model did South Korea and Taiwan follow on their path to development?
   a. The British model of colonial dominance
   b. The Chinese model of communism
   c. The German model of industrial development
   d. The Japanese model of export-led economic growth
   e. The Soviet model of collectivization

14. Which of the following pairings of East Asian countries with their approaches to economic development is INCORRECT?
   a. China has taken a modified capitalist road to development
   b. Hong Kong has one of the most *laissez*-faire economic systems in the world
   c. Japan began its development through a communist system of collectivization
   d. South Korea initiated a program of export-led economic growth
   e. Taiwan's economy is organized around small and mid-sized family firms

15. Which country of East Asia has had a long-term population control policy that has resulted in a gender imbalance?
   a. China   b. Japan   c. North Korea   d. South Korea   e. Taiwan

Answers: 1-E; 2-B; 3-A; 4-C; 5-A; 6-A; 7-C; 8-B; 9-B; 10-A; 11-C; 12-E; 13-D; 14-C; 15-A

Chapter 12
SOUTH ASIA

LEARNING OBJECTIVES
- This chapter covers South Asia, which includes India (the world's second-most populous country), Pakistan, Bangladesh, Nepal, Bhutan, Sri Lanka, and the Maldives.
- The student should understand the unique climatological challenges of this region, which include monsoons and hurricanes, as well as the accompanying flooding.
- The student should understand the challenges of feeding a large and growing population.
- Upon completion of this chapter, students should be familiar with the physical, demographic, cultural, political, and economic characteristics of South Asia.
- In addition, the student should understand the following concepts and models:

  · Monsoon
  · Green Revolution
  · Subcontinent
  · Orographic rainfall
  · Indian diaspora
  · Hinduism and the caste system

CHAPTER OUTLINE
I. Introduction
South Asia is the world's second most populous region with about 1.45 billion people. The population in this region is continuing to grow, and there are concerns about the ability of food production to keep pace. South Asia experienced British colonialism for several centuries. Since achieving independence in 1947, the two largest countries in the region, India and Pakistan, have been embroiled in conflict. The fact that these countries possess nuclear weapons is a serious concern not just in the region, but globally as well. Economically, this region is one of the world's poorest; it is not well connected with the global economy. South Asia is often called the Indian subcontinent

II. Environmental Geography: Diverse Landscapes, from Tropical Islands to Mountain Rim
  A. Large Region, with Incredible Diversity (Fig. 12.1)
  B. Building the Quadrilateral Highway
   1. Multilane highways are rare in India, with speeds slowing to 6.8mph at points from Kolkata (Calcutta) to Mumbai (Bombay)
   2. Higher taxes and tolls, destruction of homes, and religious and environmental conflicts have resulted
   3. Quadrilateral highway will connect New Delhi, Kolkata, Chennai (Madras), and Mumbai almost completed by March 2007

C. Environmental Issues in South Asia: high populations, pollution associated with early stages of industrialization, and environmental disasters contribute to environmental problems
1. Natural hazards in Bangladesh: rapid population growth has caused people to move to regions affected by monsoon-related floods, putting them in harm's way; a 2007 flood in the region affected 35 million people
2. Forests and Deforestation: vast areas in this region have been deforested to make room for agriculture, cities, or industries; many villages have a shortage of fuel wood for household cooking and use cow dung instead, which takes fertilizer that could otherwise have been used to help grow fuel crops
3. Wildlife: Extinction and Protection: South Asia is home to such endangered species as tigers and elephants; South Asia has the best environmental protection in Asia; conversion of wildlands to farmland is a major cause of habitat loss and wildlife decline
D. The Four Subregions of South Asia
1. Mountains of the North: dominated by Himalayas, Karakorum, and Arakan Yoma mountains; this region was created by the collision of two tectonic plates
2. Indus-Ganges-Brahmaputra River Lowlands: created by two major river systems that have carried sediments from eroded mountains and built vast alluvial plains; soil here is fertile and easily farmed; this region is densely settled; Indus River is

the largest in the region; the Ganges River is a transportation corridor and is also considered sacred
3. Peninsular India: this is the elevated Deccan Plateau; soils are not especially fertile except in Maharashtra state, where basaltic lava flows have produced rich soils; water supply is a problem; proposed dams threaten to displace more than 100,000 people; the Ghat Mountains (western and eastern) are in this region
4. The Southern Islands: Sri Lanka has a tropical climate and picturesque mountains; the Maldives is a chain of 1,200-plus islands with a combined land area of about 116 square miles; the islands are flat, low coral atolls
E. South Asia's Monsoon Climates (Fig. 12.7)
1. *Monsoon:* the distinct seasonal change between wet and dry periods, generally accompanied by wind; South Asia has a warm, rainy summer monsoon (June–October) and a cold, dry winter monsoon (November–February)
2. The presence of the mountains results in *orographic rainfall* associated with monsoons in this region; *orographic precipitation:* precipitation (rain, snow, sleet, fog, dew, etc.) that is dependent on the existence of a mountain range; in orographic precipitation, (1) breezes blowing from the ocean to the land carry moisture; (2) as the moist air rises up the mountainside, it cools and releases precipitation; (3) eventually, all the moisture is

released; (4) the breeze—now dry—continues over the mountain; (5) the sheltered or lee side of the mountain is usually very dry; this dry area is called a *rain shadow* and experiences a *rain shadow effect*

  F. Global Warming and South Asia
    1. Rising sea-level will be a problem for Bangladesh and small islands in the region
    2. Retreat of Himalayan glaciers threatens dry-season water supplies for agriculture
    3. India signed Kyoto Protocol, but is exempt from the main requirements because it is an LDC
    4. India signed an agreement with Japan for energy efficient technology and is working toward greater energy efficiency

III. Population and Settlement: The Demographic Dilemma (Fig. 12.11)

  A. India has more than a billion people; Pakistan has almost 170 million; and Bangladesh has 149 million; many people are malnourished
  B. The Geography of Family Planning
    1. India has had family planning policies in place for 40 years; total fertility rate dropped from 6 in the 1950s to an average of 2.9 today; in places where women are better educated, birth rates are lower; sterilization is an important family planning strategy; preference for male children causes problems similar to those in China
    2. Pakistan government's official position is that the birthrate is too high (TFR=4.1); in 2005, Pakistan urged Muslim scholars to promote family planning, but the program has met resistance.
    3. Bangladesh has made significant progress in family planning; its TFR dropped from 6.3 in 1975 to 3.0 by 2006; government strongly supports family planning and has implemented a woman-to-woman information-sharing strategy that is proving to be effective; there is no coercion; (both Pakistan and Bangladesh are Muslim)
  C. Migration and the Settlement Landscape (Fig. 12.12)
    1. South Asia is largely rural; however, migration to cities is growing as mechanization of agriculture causes unemployment or underemployment
    2. Four areas within South Asia are experiencing major out-migration: Bangladesh, the Indian states of Bihar and Rajasthan, and the northern part of India's Andhra Pradesh; most out-migrants go to large cities, but migrants from Bangladesh tend to move to rural portions of neighboring Indian states, and Nepalese migrants are moving to formerly malaria-infested lowlands along the Indian border
  D. Agricultural Regions and Activities
    1. Historically, South Asia has been relatively unproductive; since the 1970s, the *Green Revolution* (agricultural cultivation techniques based on hybrid crops and the heavy use of industrial fertilizers and chemical pesticides) has changed this
    2. Crop Zones: crops grown in South Asia include rice, wheat, oil seeds, coconuts, jute; the Punjab is India's breadbasket

3. Livestock: most South Asians do not get enough protein; meat consumption is low; India has the world's largest cattle population, and milk is an important protein source; cows are sacred; *a white revolution* has increase dairy efficiency in recent decades

4. The Green Revolution: began in the 1960s, introduced hybrid crops to increase food production; the Green Revolution has succeeded in increasing food production but some say it has been an ecological and social failure; Green Revolution practices require large amounts of industrial fertilizer and frequent pesticide applications that are expensive and polluting; only more prosperous farmers can afford the new seed strains, irrigation equipment, farm machinery, fertilizers, and pesticides to support this technology

5. Future Food Supply: can there be additional improvements in productivity, or has the Green Revolution done all it can do? Additional irrigation is possible, but *salinization* (buildup of salts in farm fields) is already a problem, and water tables are falling; better transportation can improve food distribution

E. Urban South Asia

1. Only about 30-40% live in cities, but this region contains some of the largest cities in the world; rapid growth of cities has led to problems of homelessness, poverty, congestion, water shortages, air pollution, sewage disposal problems, and squatter settlements (called *bustees* in this region)

2. Mumbai (Bombay), India: largest city in South Asia with 16 million, India's financial, industrial, and commercial center, and the major port on the Arabian Sea; this city has enormous contrasts in wealth and poverty, but the city is relatively orderly and crime-free

3. Delhi, India: India's capital city, with more than 11 million people; old Delhi, a former Muslim capital, is a congested town of tight neighborhoods; New Delhi has wide boulevards, monuments, parks, and expansive residential areas; it began as a British colonial capital; Delhi has some of the worst air pollution in the world

4. Kolkata (Calcutta): population exceeds 12 million; has problems typical of rapidly growing cities in developing countries: problems with water, sewage, power, flooding during the wet season, pollution, and congestion; industry is in decline here; Calcutta is culturally vibrant and has fine schools, theaters, and publishing firms

5. Dhaka, is the capital and major city of Bangladesh with the country's largest industrial concentration; it's a global center for clothing, shoe, and sports equipment manufacturing

6. Karachi, Pakistan, is a port city with more than 7 million people; Karachi was the capital of Pakistan till 1963; it is Pakistan's largest urban area and its commercial core; Karachi has serious political and ethnic

tensions that have sometimes been violent

7. Islamabad, Pakistan is a forward capital, designed to reaffirm Pakistan's political interest in Kashmir

IV. Cultural Coherence and Diversity: A Common Heritage Undermined by Religious Rivalries

A. Historically, South Asia is a well-defined cultural region
1. 1,000 years ago, Hinduism united the region; today, religious differences are creating challenges
2. Since the 1980s, there has been a growth in *Hindu nationalism;* proponents of this movement promote Hindu values as the essential and exclusive fabric of Indian society; this group now has considerable power both at the federal level and in many Indian states, through the Bharatiya Janata Party (BJP); conflicts between this group and Muslims has been an issue
3. Islamic fundamentalism is also an issue, especially in Pakistan, which shares a border with Afghanistan , home of the ultra-fundamentalist Taliban

B. Origins of South Asian Civilizations
1. Indus Valley (Pakistan), was the site of South Asia's earliest origins some 4,000 years ago; newer settlements (around 800 BCE) followed in the Ganges Valley
2. Hindu Civilization: Hinduism is tremendously complicated faith that incorporates diverse forms of worship and lacks an orthodox creed; Hindus recognize various deities, but all are manifestations of a single divine entity; *Sanskrit* is the sacred language of Hinduism; Hindus believe in the transmigration of souls from being to being through reincarnation; Hinduism has a *caste system* (the strict division of society into different hierarchically ranked hereditary groups); recent research suggests that Hinduism and the caste system emerged through gradual social and cultural processes
3. Buddhism: Buddhism challenged the caste system from within, founded by Prince Siddhartha Gautama (the Buddha), born in 563 B.C.E.; Buddha rejected the life of wealth and power that was his birthright and sought enlightenment or *nirvana* (mystical union with the cosmos); he preached that the path to nirvana was open to anyone, regardless of their social class; Buddhism spread under the Mauryan Empire of South Asia, but ultimately did not replace Hinduism in India; Buddhism has expanded through most of East, Southeast, and Central Asia
4. Arrival of Islam: Arab invaders and later Turkish-speaking Muslims settled in the region and brought Islam to the region; during the 16th and 17th centuries, India's Islamic Mughal Empire dominated; Hindus from the lower castes were especially attracted to Islam, which offered them an escape from Hinduism's rigid social order; although the religions are very different, Hindus and Muslims coexisted amicably until the 20th century

5. The Caste System: the caste system is not uniformly distributed across South Asia, The main caste groups: the word *caste* has a Portuguese origin and combines two distinct local concepts: *varna* (the ancient fourfold hierarchy of the Hindu world) and *jati* (the hundreds of local groups that exist at each varna level); both varna and jati are hierarchically ordered, but the exact order of precedence is not entirely straightforward. The changing caste system: The caste system is being deemphasized, especially among better-educated people; affirmative action programs in place for members of lower castes

C. Contemporary Geographies of Religion (12.21)

1. Hinduism: the vast majority of Indians are Hindu; fewer than 1 percent of Pakistanis are Hindu; Hinduism a minority religion in Bangladesh and Sri Lanka

2. Islam: Bangladesh and Pakistan are overwhelmingly Muslim; 15 percent of Indians are Muslim (about 150 million—more than the total population of any Muslim heartland of Southwest Asia and North Africa); Muslims are found throughout India; Sri Lanka is about 9 percent Muslim; Maldives nearly all Muslim

3. Sikhism: originated in the late 1400s in the Punjab, near the modern boundary of India and Pakistan; Sikhism combines elements of Hinduism and Islam; periodic persecution of Sikhs bred militarism, and they carved a kingdom for themselves in the early 1800s; the Punjab is 60

percent Sikh; Sikh men are readily identifiable because they do not cut their hair or beards and instead wear their hair wrapped in a turban and often tie their beards close to their faces

4. Buddhism and Jainism: Among Sri Lanka's dominant Singhalese people, Theravada Buddhism has become a virtual national religion; Tibetan Buddhism (or Lamaism) survived as the majority religion in the high valleys of the Himalayas and in Bhutan and a portion of Kashmir; Jainism emerged as a protest against orthodox Hinduism; Jainism stresses nonviolence; because plowing can kill small creatures, Jains are not permitted to farm; most Jains have turned to trade for their livelihood, and many have become wealthy and constitute one of the wealthiest groups in India; Jains are concentrated in northwest India

5. Other religious groups: Parsis or Zoroastians (concentrated around Mumbai) are very prosperous; the Parsis arrived as refugees who fled Iran after the arrival of Islam in the seventh century; Zoroastrianism focuses on the cosmic struggle between good and evil; several of India's largest industrial firms are controlled by Parsi families; Indian Christians are more numerous than either Jains or Parsis; Christianity arrived around A.D. 200; most are affiliated with the Syrian Christian Church of Southwest Asia; in Goa, a former Portuguese colony, Roman Catholics and Hindus each

comprise about half the population of this state; there are a small number of Jews in the region

D. Geographies of Language (Fig. 12.24)
1. India contains an important linguistic dividing line; north of the line, languages are in the Indo-European family; south of the line, languages are Dravidian
2. The Indo-European North: Iranian languages (e.g., Baluchi and Pashtun) are found in western Pakistan; Indo-Aryan languages are all closely related, and Hindi is the most widely spoken language of South Asia, with Bengali second; Urdu is the official language of Pakistan
3. Languages of the South (Dravidian): four main Dravidian languages each have a close association with specific Indian states (Kannada in Karnataka; Malayalam in Kerala; Telugu in Andhra Pradesh; Tamil in Tamil Nadu); Singhalese is the dominant language of Sri Lanka
4. Linguistic Dilemmas: India has 23 official languages); Hindi is spreading through popular media and educational TV; *Linguistic nationalism*: linking of a specific language with political goals; could there be a national language for India?
5. The Role of English: the main integrating language of India, and is widely used in Pakistan, Bangladesh, and Sri Lanka

E. South Asians in a Global Cultural Context
1. Widespread use of English facilitates the spread of global culture into the region
2. English also enables South Asians' cultural works to reach global audience: Rabindranath Tagore won Nobel Prize for literature in 1913; Salman Rushdie and Vikram Seth became well-known as authors in 1980s and 1990s
3. Emigration of many South Asians spreading their culture around the world: South Asians (mostly Pakistanis) migrating to Britain; South Asians (mostly Indians) migrating to the United States (Fig. 12.26)

V. **Geopolitical Framework: A Deeply Divided World Region** (Fig. 11.28)
A. British Rule Politically United This Region in Mid-1800s: any prior unity was cultural; independence from Britain resulted in separation of Pakistan from India
B. South Asia Before and After Independence in 1947
1. British arrived in 1500s, settled on coasts, while the Mughal Empire (Muslims) ruled the north, and Indians ruled the south; during the 1700s, the Mughals weakened and the Hindu state became stronger
2. The British Conquest: the British East India Company, a private organization acting as an arm of the British government, monopolized trade in India and exploited political chaos of the region to stake out its empire, eventually controlling most of the region
3. From Company Control to British Colony: the reduction in size of indigenous Indian states, along with the growing arrogance of British officials, led to an

135

uprising in 1856 across much of South Asia; Britain crushed the rebellion, and then ruled this region directly; Britain tried to interfere with Nepal and Bhutan (which nominally were independent), but suffered defeat at the hands of the Afghans, who retained their independence

4. Independence and Partition, 1947: independence movements began in the region in the 1920s; Britain withdrew from South Asia in 1947; the father figure of India favored a single unified state at this point, but Muslim leaders argued for two new countries (Hindu-majority India, and Muslim-majority Pakistan); the two-country partition won out; mass migrations and many deaths followed the partitioning; Pakistan has remained relatively unstable politically and allowed almost full autonomy to the Pashtun tribes in the northwest; this region lent much support to Osama bin Laden's al-Qaeda organization

5. Geopolitical Structure of India: India is organized as a *federal state,* with significant power vested in individual states; Indian states organized according to linguistic geography, but only the largest ethnic groups had their own states; smaller groups are also pushing for their own states

C. Ethnic Conflicts in South Asia
1. Kashmir has a large Muslim core, with an Indian district and a Buddhist district; some Kashmiri Muslims would like to join Pakistan; India is firm in its desire to keep Kashmir; some Kashmiris want their own

separate state; there is no end in sight to this conflict

2. The Punjab is a place of conflict between Sikhs and Hindus; Hindus consider Sikhism as a sect of Hinduism; radical Sikhs want independence from India; the Sikhs' militarism is an issue, and in 1984, a Sikh bodyguard of President Indira Ghandi assassinated her; Sihks now under martial law

3. The Northeastern Fringe: this remote upland area lies in India's extreme northeast; migrants to this region from Bangladesh and other Indian provinces are viewed negatively by the indigenous peoples; locals have attacked the immigrants, because they consider the immigrants a threat to both their land and their cultural integrity

4. Sri Lanka: religious and linguistic differences fuel ethnic conflict; in this case, the majority Buddhist Sinhalese favor a unitary government, with some arguing that Sri Lanka should be a Buddhist state; the minority Hindu Tamils support political and cultural autonomy; war erupted in 1983 when the Tamil Tigers attacked the Sri Lankan army; a solution is unlikely in the near term, in spite of a 2003 ceasefire

5. The Maoist Challenge: Minor insurgencies in the Indian states of Bihar, Jharkhand, Chattisgarh, and Andhra Predesh in 2007 were inspired by China's Mao Zedong; a Maoist insurgency in Nepal remains a source of conflict

D. International and Global Geopolitics
  1. Cold war between India and Pakistan has continued; the countries regard each other as enemies, and both have nuclear weapons; during the Cold War, Pakistan was an ally of the United States, while India remained neutral, with slight leanings toward the U.S.S.R.; Pakistan also has built an alliance with China, from which it has obtained sophisticated military equipment; China and India are engaged in a territorial dispute
  2. Pakistan after 9/11: has agreed to help the U.S. in the war against terrorism, especially in Afghanistan, which is Pakistan's next-door neighbor; relationship has become more complex since the attacks on the World Trade Center on September 11, 2001; before then, Pakistan supported the Taliban, while India provided some help to the Northern Alliance; but since then, with Pakistan's approval, the United States has set up military bases in Afghanistan; Islamic fundamentalists in Pakistan are unhappy, and there is a great deal of anti-American sentiment in Pakistan.
  3. India's Geopolitical Fears and Ambitions: Bangladesh and India are generally cordial; India supported Bangladesh's independence from Pakistan; India would like to assume a position as the dominant regional power in South Asia and the India Ocean Basin, but ongoing conflicts (both domestic and with Pakistan) and economic weakness make this a long shot

VI. Economic and Social Development: Burdened by Poverty
  A. South Asia is a land of contradictions. Along with Sub-Saharan Africa, it is the poorest world region. South Asia has a growing middle class, but many social groups are virtually cut off from the processes of development. South Asia has made some world-class scientific and technological achievements, but its illiteracy rate is among the world's highest; while South Asia's high-tech businesses are well integrated into the global economy, as a whole the region's economy is self-contained and inward-looking
  B. South Asian Poverty
    1. More than 300 million (about 30 percent) Indians live below that country's poverty line, but there is a growing middle class
    2. Bangladesh is in even worse shape: 30 percent of its people are undernourished; in India, 20% are undernourished and in Pakistan, 25% are undernourished
    3. Many South Asian children work in sweatshops
    4. Many people are homeless
  C. Geographies of Economic Development
    1. The Himalayan Countries: Nepal and Bhutan largely subsistence-oriented; Bhutan is purposely disconnected from the world economy and has only recently permitted tourists to enter, but exports hydroelectricity to India; Nepal is more integrated with Indian and world economies, its tourist industry brings income, but also potential ecological damage
    2. Bangladesh: most people involved in market economies of commercial export crops (rice and jute); poverty is widespread

and extreme; environmental degradation exacerbates poverty; Bangladesh's low wages make it internationally competitive in textile and clothing manufacturing; the Grameen Bank gives low-interest loans to poor women for small scale businesses

3. Sri Lanka and the Maldives: measured by conventional criteria, Sri Lanka's economy is the second most developed in the region; its primary exports are textiles, rubber, and tea; by global standards, Sri Lanka remains poor, and any progress is hampered by its ongoing civil war; the Maldives is the most prosperous South Asian country based on per capita GNI, but its economy and population are small; fishing and international tourism are the major industries

4. Pakistan: has reasonably well-developed urban infrastructure; Pakistan's purchasing power parity and GNP are higher than either Bangladesh or India; agriculture is productive and the cotton-based textile industry is large and relatively successful, but growth potential may be limited; Pakistan is burdened by high military spending; the best farmlands are controlled by a small, powerful landlord class that pays no taxes

5. India's Less-Developed areas: India's economy is the largest in the region; it is more prosperous in the west, less prosperous in the east, with a great deal of variation from state to state

6. India's Centers of Economic Growth: *Punjab and Haryana* are showcases of the Green Revolution, and their economies are based on agriculture; *Gujarat* and *Maharashtra* are noted for their industrial and financial clout and agricultural productivity; people from Gujarat are disproportionally represented in the *Indian diaspora* (the migration of large numbers of Indians to foreign countries), and their remittances bolster the local economy; Maharashtra viewed as India's economic pacesetter; Mumbai is a financial and film center and its port handles 40 percent of India's total trade; India's fast-growing high-tech center is in *Bengaluru (Bangalore)* and *Karnataka;* India has been especially competitive in software development

D. Globalization and India's Economic Future (Fig. 12.40)
1. Indian economic system is a mix of socialism and capitalism, with widespread private ownership, but government control of planning, resource allocation, and certain heavy industries
2. High trade barriers protect the economy from global competition
3. Liberalization of economy began in early 1990s: economy was opened to imports and multinational firms
4. Cheap goods from China threaten India's growth; low investment in infrastructure in India has also hindered development

E. Social Development
1. Overall low levels of health and well-being exist in South Asia; there are also geographical

patterns, with better health and longer life expectancy and education levels in western India; the state of Bihar is consistently on the bottom, while Punjab is consistently on top

2. The Educated South: Sri Lanka has higher health and education indicators than would be expected of a developing country; Kerala (southwestern India) has best indicators of social development in India

3. The Status of Women: women in the Hindu tradition are forbidden from inheriting land; widows may not remarry; many women leave their families shortly after puberty to join their husband's families; women still suffer discrimination: they have lower rates of literacy, and suffer from *differential neglect,* which occurs when children of one sex (usually girls) receive poorer nutrition and health care than the children of the other sex (usually boys); the excess of boys in India suggests that many boys receive better treatment; reasons for differential neglect of girls: boys remain with family even into adulthood, contributing to family wealth, while girls go to live with their husband's families; families must provide dowry for their daughters' prospective husbands; evidence suggests that the social position of women is improving

VII. Conclusion

South Asia's size and growth rate have assured it a prominent role in discussions of world problems and issues. It is poised to become the most populous region in the coming century, and feeding this burgeoning population will pose a major challenge for this region. The geopolitics of the region cause much international concern, particularly since two rivals in this region, India and Pakistan, have nuclear weapons. Other ethnic conflicts are found throughout the region. Economic development in South Asia is highly variable. While one Indian state is developing a strong computer hardware and software industry, other states are among the worlds poorest. Bangladesh is developing a strong textile industry, while Nepal is opening its doors to tourists. Women in South Asia continue to face many challenges, but there are bright spots, such as the efforts of Grameen Bank to help women establish small-scale enterprises in Bangladesh. Much uncertainty remains about the direction of this region in the coming decades.

PRACTICE MULTIPLE CHOICE QUIZ

1. What are the tallest mountains in South Asia --- and the world?
    a. Eastern Ghats
    b. Himalayas
    c. Satpura Range
    d. Vindhya Range
    e. Western Ghats

2. Why is Bangladesh so vulnerable to the cyclones and wet monsoons of South Asia?
   a. Concentration of the agricultural activity of Bangladesh in the fertile delta valley
   b. Continued population growth in the region
   c. Deforestation of the headwaters of the Ganges and Brahmaputra Rivers
   d. High density and clustering of the population of Bangladesh on the low-lying delta area
   e. All of the above

3. Which country of South Asia protects its wildlife better than most other countries of Asia?
   a. Bhutan   b. Bangladesh   c. Pakistan   d. India   e. Nepal

4. All of the following statements about monsoons are correct, EXCEPT....
   a. Monsoons are the dominant climatic force in most of South Asia
   b. Monsoons are caused by large-scale meteorological processes
   c. Monsoon is the Hindi world for El Niño
   d. A monsoon is a distinct seasonal change of wind direction
   e. A monsoon can be wet or dry

5. Which of the following statements is TRUE?
   a. South Asia is the most populous region in the world
   b. South Asia is growing at a slower rate than East Asia
   c. A cultural preference for males in South Asia complicates family planning efforts
   d. India has the highest population in the world
   e. All of the above

6. Which country of South Asia has made significant strides in family planning by employing a large number of women as fieldworkers who take information to other women in the country's villages?
   a. Bangladesh   b. India   c. Nepal   d. Pakistan   e. Sri Lanka

7. Which of the following statements about family planning in India is INCORRECT?
   a. Especially in northern India, a lower fertility rate is accompanied by a higher ratio of male to female infants
   b. India's population is no longer growing
   c. Measures taken to slow India's population growth have caused a drop in the total fertility rate from 6 in the 1950s to its current rate of 2.9
   d. In places where women's literacy has increased most dramatically, the birthrate has declined rapidly
   e. Widespread concern over India's population growth began in the 1960s

8. Where do most of the people of South Asia live?
   a. In the far north highland
   b. In the cities
   c. In the arid lands of the northwest
   d. In compact rural villages and small towns
   e. In apartments

9. Why do so many South Asians receive inadequate protein?
   a. Poverty
   b. Most Hindus are vegetarians
   c. Meat is expensive
   d. Cattle are sacred in Hinduism
   e. All of the above

10. Which of the following pairings of place with religion in South Asia is INCORRECT?
    a. Sri Lanka – Buddhism
    b. Pakistan – Islam
    c. Nepal – Jainism
    d. India – Hinduism
    e. Bangladesh – Islam

11. All of the following statements about English in India are true, EXCEPT...
    a. English is the main integrating language of India
    b. Indian children are taught English early, regardless of their social and economic class
    c. Information technology companies in India sometimes ask their employees to watch American TV shows in order to gain fluency in American pronunciation and slang
    d. Previously, many educated Indians learned English because of its political and economic benefits under British colonialism
    e. Use of English carries substantial international benefits

12. What was the main reason for partitioning South Asia into three countries (India, Pakistan, Bangladesh) after it achieved its freedom from Britain?
    a. Ethnic rivalries had erupted into violence and there was no other choice
    b. Linguistic differences made unity impossible
    c. Mountains separated these regions, so it was decided to create a physiographic boundary
    d. Muslims in the northeast and northwest parts of the region did not wish to live under Hindu rulers
    e. The boundaries were a legacy of British colonization, much as the Berlin Conference determined the borders of today's African countries

13. Which country of South Asia has nuclear weapons?
    a. India
    b. Pakistan
    c. Bangladesh
    d. A and B above
    e. A and C above

14. Which of the following statements about South Asia is true?
    a. Some South Asians have immense fortunes
    b. South Asia has a sizable and growing middle class
    c. South Asia has some of the highest illiteracy rates
    d. South Asia is (along with Sub-Saharan Africa) among the poorest world regions
    e. All of the above

15. Why does Sri Lanka have the highest life expectancy and literacy rates in South Asia?
    a. It has a large software development industry
    b. It has established universal primary education and inexpensive medical clinics
    c. It has received a great deal of foreign direct investment
    d. It is a very rich country
    e. It receives large amounts of foreign aid

Answers: 1-B; 2-E; 3-D; 4-C; 5-C; 6-A; 7-B; 8-D; 9-E; 10-C; 11-B; 12-D; 13-D; 14-E; 15-B

Chapter 13
SOUTHEAST ASIA

LEARNING OBJECTIVES
- This chapter covers Southeast Asia, which includes Burma (Myanmar), Thailand, Laos, Cambodia, Vietnam, and insular (island) Southeast Asia, which includes Indonesia, the Philippines, Malaysia, Brunei, and Singapore.
- The student should understand the unique biogeography associated with islands.
- The student should also understand the idea of an export-based economy and the ways that such an economy fits in with the global economy.
- Upon completion of this chapter, students should be familiar with the physical, demographic, cultural, political, and economic characteristics of Southeast Asia.
- In addition, the student should understand the following concepts and models:

  - Typhoon
  - Swidden
  - Transmigration
  - Domino Theory
  - ASEAN (Association of Southeast Asian Nations)

CHAPTER OUTLINE

I. Introduction

Southeast Asia is an excellent example of the promise and perils of late 20th century globalization. Thoroughly plugged into global capitalism, this region experienced rapid economic growth and development in the 1980s, only to fall victim to a bust cycle in the middle and late 1990s. Southeast Asia is no stranger to the outside world, having experienced colonial dominance and expansionist imperialism for many years. Troubled not only by economic challenges, internal ethnic and social conflicts are also part of the mix in Southeast Asia. In spite of all these challenges, the countries of this region have forged an important union, the Association of Southeast Asian Nations (ASEAN), which offers great promise for promoting the self-determination of the region.

II. Environmental Geography: A Once-Forested Region (Fig. 13.3)
A. The Tragedy of the Karen
  1. The Karen, a tribal people in Burma, had never been fully integrated into the Burmese

kingdom; when the British colonized Burma, 30% of the Karen converted to Christianity. As a result of this relationship, the British gave many Karen people positions in the colonial government. The Burmans (ethnic Burmese, and the majority in Burma) resented this favoritism. After the British left, the Karen lost their favored position, and conflict between the Karen and Burmans ensued. The Karen rebelled openly and established an insurgent state of their own. They supported their state by smuggling goods between Thailand and Burma, and by mining gemstones in their region. By the early 1990s, the Burmese army had overrun the Karen. The Burmese army was successful in part because they were able to enlist Thailand's assistance in preventing Karen soldiers from finding sanctuary in Thailand.

   2. To get this agreement, Burma agreed to allow Thailand to have access to Burma's valuable teak forests; this agreement has resulted in pressure on the forests of Burma

B. The Deforestation of Southeast Asia

   1. Globalization has had a profound effect on the environment of Southeast Asia; export-oriented logging has resulted in a loss of trees that far exceeds the loss either from indigenous peoples or by colonizers

   2. Japan was the first country to globalize world forestry in the 1960s, but other Asian countries (Taiwan, South Korea, Indonesia, Malaysia) have developed their own wood-products firms

   3. Although countries such as Indonesia see the potential associated with clearing forest lands for food production, agriculture and population growth are not the main causes of deforestation in Southeast Asia; some logged-off lands are reforested

   4. Now most Southeast Asian countries have widely publicized bans on the export of raw logs, and some (like Thailand) have banned forest logging altogether

   5. Malaysia has been the leading exporter of tropical hardwoods from Southeast Asia, with 60% of these logs going to Japan; Malaysia will be almost completely deforested in the near future

   6. Indonesia has two-thirds of the region's forest areas, and about 10% of the world's true tropical rain forest; much of it has already been lost to logging

   7. Thailand cut more than 50% of its forests between 1960 and 1980; logging has been illegal there since 1995;

damage to the landscape has been severe

C. Fires, Smoke and Air Pollution
1. Air pollution in this region is surprisingly strong
2. Sources of air pollution: (1) urban smog in the region's fast-growing cities, (2) smoke from clearing forests by burning; (3) tinder-dry forests due to drought caused by El Niño; (4) the peat bogs of Kalimantan dried out and burned for months
3. Results: unhealthy air; lung cancer and pulmonary disease in this region kill at five times the U.S. rate

D. Patterns of Physical Geography (Fig. 13.1)
1. This region is one of the world's three main zones of tropical rain forest; this is a region of diverse landforms, with a great deal of tectonic activity (both earthquakes and volcanoes)
2. Mainland Environments: mountains and lowlands dominated by rivers; mountains form the northern boundary of this region; rugged terrain has resulted in sparse settlement; this region has several large rivers (Mekong, Irrawaddy, Red River, and Chao Phraya)
3. Monsoon Climates (Fig. 13.7): almost all of mainland Southeast Asia lies in the tropical monsoon zone; characterized by a hot rainy season from May to October;

rainfall totals may exceed 100 inches per year
4. Insular (island) Environments: Southeast Asia is an archipelago (island) environment; *Indonesia* has more than 13,000 islands, tectonic activity created a mountainous spine and left volcanic peaks; Philippines has more than 7000 islands, and includes active volcanoes; this region also includes the world's largest expanse of shallow ocean, covering the *Sunda Shelf;* the rich marine life resulted in Southeast Asian peoples' adopting maritime ways of life
5. Equatorial Island Climates: island climates are more complicated than those on the mainland because of a more complex monsoon effect, the influence of Pacific *typhoons,* and the equatorial location; *typhoons* (large tropical storms, similar to hurricanes, that form in the western Pacific Ocean); the islands have very little seasonal variation in temperature and precipitation is year-round

E. Global Warming and Southeast Asia
1. Most people of Southeast Asia live in coastal areas, making them vulnerable to rising sea levels
2. Monsoon storms could strengthen
3. Most S.E. Asian countries ratified the Kyoto Protocol, but

because they are mostly LDCs, they do not need to cut back on greenhouse gasses
4. Burning and destruction of rainforests in the region contributed to greenhouse gas emissions and climate change

III. Population and Settlement: Densely Settled Lowlands amid Settled Uplands (Fig. 13.10)
A. Settlement in the region is relatively sparse (compared to East Asia and South Asia)
1. Migration to this region by sea has meant concentration of populations in deltas and coastal areas
2. Many favored lowlands have seen dramatic population growth in the past few decades
3. There has been a major rural-to-urban migration
B. Settlement and Agriculture
1. Poor rain forest soils lead to sparse settlement; plant nutrients are locked up in the vegetation, while soils lack nutrients, partly because the rain washes them away; shifting cultivation is well adapted to this environment; volcanic soils are more fertile
2. *Swidden* in the Uplands (other names for swidden are shifting agriculture and "slash and burn"): small plots of land are cut (often by hand), planted in indigenous subsistence crops; farmers use the crops till fertility drops; the land reverts to woody vegetation and eventually return to their original form; swidden is sustainable when populations are small and stable, and the land available is large enough; this is a threatened way of life; when swidden is no longer possible, former practitioners may turn to cash crops (including opium for the global market); the mountainous area of Burma where opium is grown is known as the *Golden Triangle*
3. Plantation Agriculture: began in Southeast Asia with European colonization; it is often located on coasts because of the need to transport the product overseas; typical plantation products include rice, rubber, cane sugar, tea, pineapple, and copra. In Southeast Asia, most plantation workers are indigenous peoples from the highlands or contract laborers from India or China
4. Rice in the Lowlands: grown in commercial quantities for trade and export, along the Irrawaddy River (Burma), Chao Phraya (Thailand), and Mekong (Vietnam and Cambodia)
C. Recent Demographic Change
1. Southeast Asia does not have the overwhelming population pressures of East or South Asia, so there is wide variation in relevant policies in this region

2. Population Contrasts: the Philippines have relatively high fertility and growth rates, many Roman Catholics; Laos (which is Buddhist) has a high TFR in the region (4.8), with little economic development; Thailand, with a high level of economic development, has a TFR of 1.7 in part because the government promoted family planning; Indonesia has the region's largest population (231.6 million), is still above replacement level, even though the government promotes family planning; Singapore, a city-state, has the lowest TFR in the region (1.3); Laos has high fertility rate, and low life expectancy in the region as a result of civil strife and violence

3. Growth and Migration: Indonesia had a policy of *transmigration* (relocation of people from one region to another within the same national territory) to move people from densely populated Java to the outer islands; the program has been drastically cut back; Philippines has about 8 million citizens overseas (many as temporary guest workers)

D. Urban Settlement
1. The overall urbanization rate in Southeast Asia is still relatively low, but cities in the region are growing rapidly

2. *Primate cities* (single, large urban settlements that overshadow all others) are common in Southeast Asia; examples include Bangkok, Manila, Jakarta

3. Thailand, Philippines, and Indonesia are all trying to encourage growth of secondary cities by decentralizing economic functions

4. Kuala Lumpur, Malaysia, has received substantial government investment, and has become a modern city of grandiose ambitions; when completed in 1996, Kuala Lampur's Petronas Towers was the world's tallest building

5. Singapore is a city-state, where space is at a premium; Singapore has no squatter settlements, and most of its buildings are new, mostly high-rise

IV. Cultural Coherence and Diversity: A Meeting Ground of World Cultures
A. Southeast Asia has never been dominated by a single culture, but instead has been a meeting ground for cultures from many regions
B. The Introduction and Spread of Major Cultural Traditions
1. South Asian Influences include Hindus from India who came 2000 years ago to Burma, Thailand, Cambodia, Vietnam, Malaysia, and Indonesia; a second wave of

South Asian religious influence came in the form of Theravada Buddhism in the 13th century to Burma, Thailand, Laos, and Cambodia

2. Chinese Influences: Vietnam was a province of China until around 1000; the presence of Mahayana Buddhism and Confucianism are a reminder; during the 19th and 20th centuries, many Chinese (mostly men) migrated to Southeast Asia; some married local women and established mixcd communitics; in thc 19th century, Chinese women began to migrate in large numbers, resulting in ethnically distinct Chinese settlements; in many places in Southeast Asia, relationships between Chinese minority and indigenous majority are strained, partly because many Chinese maintain their Chinese citizenship, because they are relatively prosperous, or because they dominate local economic activity

3. The Arrival of Islam: occurred around 1200 A.D.; by 1650, Islam had mostly replaced Hinduism in Malaysia and Indonesia (which is the world's most populous Muslim country; 87% of its 200 million people are Muslim); the practice of Islam varies from country to country within the region,

from relaxed practice to rigid fundamentalism

4. Christianity and Tribal Cultures: European missionaries came to Southeast Asia in the late 19th and early 20th centuries; Catholicism has taken root in Vietnam (under the French) and in the Philippines (under the Spanish); converts to Christianity were more likely to have been practitioners of *animist religions* (these focus worship on nature's spirits and ancestors), rather than Hindus, Buddhists, or Muslims

5. Religion and Communism: Vietnam, Cambodia, and Laos adopted communism by 1975, and discouraged religious practice; today Vietnam's government is struggling against revival of faith in the country

C. Geography of Language and Ethnicity (Fig. 13.21): linguistic influences are many and complex

1. The Austronesian Languages: probably originated prehistorically in Taiwan; today most insular (island) languages in Southeast Asia are in this family; the most common language in this group is Malay, which is the *lingua franca* (an agreed-upon common language to facilitate communication) of the insular realm; other languages in the group are Indonesian, Javanese,

Balinese, Sundanese, and Tagalog
2. Tibeto-Burman Languages include Burmese and Tai-Kadai Languages, originating in Southern China
3. Tai-Kadai languages are found in Thailand and Laos; most of these are spoken in small local areas (sometimes isolated)
4. Mon-Khmer languages: probably once covered virtually all of mainland Southeast Asia, and involve two major languages: Vietnamese (the national language of Vietnam), and Khmer (the national language of Cambodia)
D. Southeast Asian Culture in Global Context
1. European colonization brought changes in economic and education systems; when Southeast Asian people won their independence, some tried to isolate themselves from the global system (e.g., Burma), while others were more receptive to outside influence (e.g., Philippines), and still others criticize Western influences (e.g., Malaysia and Singapore)
2. English, the global language causes ambivalence: it is at once the language of questionable popular culture, and also the language of international business and politics

3. Nationalists in various countries promote their own languages
4. Singapore encourages the use of Mandarin Chinese

V. **Geopolitical Framework: War, Ethnic Strife, and Regional Cooperation (Fig. 13.23)**
A. ASEAN gives Southeast Asia a geopolitical regional coherence, but many states still struggle with serious ethnic and regional tension; in 1999, Indonesia gave up control of the island of Timor
B. Before European Colonialism
1. All the modern countries of mainland Southeast Asia existed before European colonization: Cambodia came into being in the 12th century; Burma, Siam (Thailand), Vietnam (Annam) by the 1300s
2. On the islands of Southeast Asia, most societies were organized at the village level
C. The Colonial Era
1. Portuguese arrived first around 1500, followed by the Spanish in the Philippines; then came the Dutch (1600s); their goal was to trade in the exotic goods of the region
2. Dutch power: By 1700, the Dutch dominated the region
3. British, French, and U.S. Expansion: the colonial relationships were these: (1) British in Burma and Malaysia; (2) French in Laos, Cambodia, Vietnam; (3) Dutch in Indonesia; (4) Portuguese in East Timor; (5)

Spanish, and later the United States, in Philippines

4. Growing Nationalism: organized resistance to occupation began in the 1920s; after WWII, most Southeast Asian countries were granted their independence

D. The Vietnam War and Its Aftermath

1. French occupied Vietnam after WWII

2. A communist resistance movement arose, led by Ho Chi Minh; France allowed this group to establish a separatist government in the north; France fought against North Vietnam for 10 years, then withdrew in defeat

3. An international peace council divided Vietnam into two (like Korea): North Vietnam was allied with U.S.S.R. and China; South Vietnam was allied with the United States

4. Fighting continued, as North Vietnam sought to reunite Vietnam

5. Laos and Cambodia became involved: communist Pathet Lao challenged the central government of Laos; communist *Khmer Rouge* gained influence in Cambodia

6. U.S. Intervention: United States began sending advisors to the region in the early 1960s, as *The Domino Theory* provided the rationale (if one country went

communist, others would also become communist—just like dominoes); the logical conclusion of this theory is that eventually all of Southeast Asia would be part of the Soviet Union-Communist China world

7. U.S. began bombing in 1964, and began sending troops in 1969; by 1969, half-a-million U.S. soldiers were in Vietnam, and casualties were high; U.S. troops began to withdraw in the early 1970s

8. Communist Victory: Saigon, Vietnam, fell in 1975, and Vietnam reunited in 1976; many people fled the country

9. In Cambodia, the Khmer Rouge installed one of the world's most brutal regimes ever, led by Pol Pot; city dwellers were forced into the countryside to become peasants, wealthy and highly educated people were executed; millions of people lost their lives

10. Laos established a new communist government, it is not a democracy and its development remains at a low level; some Laotian immigrants in the U.S. oppose the current government

E. Geopolitical Tensions in Contemporary Southeast Asia

1. Conflicts in Indonesia East Timor: Began in 1975, when Portugal withdrew from this area, people fought for freedom against brutal

Indonesian government and were successful in gaining independence in 1999.

Western New Guinea (Papua): Indigenous peoples rebelled against influx of Javanese migrants and mining and lumber firms; this region has been renamed "Papua," but conflict remains, and there is still desire for independence.

Ethnic conflicts: The Aceh region of Sumatra had demanded the creation of an independent Islamic state; the devastation caused by the December 2004 tsunami brought the Islamic government and Aceh to a settlement

2. Regional Tensions in the Philippines: the Islamic southwest presents the greatest secessionist challenge

3. The Quagmire of Burma: Burma is the most war-ravaged country of Southeast Asia; most of the people who are rebelling want to maintain their cultural traditions, lands, and resources (e.g., the Karen); many insurgencies have been supported by opium and heroin trade

F. International Dimensions of Southeast Asian Geopolitics

1. Conflicts in the region have been reduced with the creation of ASEAN (Association of Southeast Asian Nations)

2. Territorial Conflicts: (1) Philippines claim Malaysian state of Sabah; (2) Malaysia interested in Malay-speaking, Muslim region of Thailand; the Spratly and Paracel islands are subject of territorial disputes by the following countries: Philippines, Malaysia, Vietnam, China, and Taiwan; concern about China from the countries of Southeast Asia provides motivation for ASEAN unity; ASEAN's current mission is to facilitate regional political cooperation in order to control (rather than be controlled by) external global forces

3. Global Terrorism and International Relations: rise of radical Islamic fundamentalism, terrorist attacks, but these groups have very little support among people in the region

VI. Economic and Social Development: The Roller-Coaster Ride of Globalized Development

A. Until the late 1990s, the economic progress in Southeast Asia was considered a model of a new globalized capitalism

1. Investment capital came from Japan, the U.S., and then from international investment portfolios

2. Some ASEAN countries have established free-trade agreements among themselves

B. Uneven Economic Development

1. The Philippine Decline: in the 1960s, the Philippines

were the most highly developed country in Southeast Asia; since the 1980s, the Philippines' economy has been in decline; there are several reasons for the decline: (1) corruption; (2) *crony capitalism* (occurs when the leader's friends are granted huge sectors of the economy, while those perceived as enemies have their property expropriated); (3) *kleptocracy:* government of thieves; (4) as Philippines fell into downward economic spiral, wealthy Filipinos invested their money elsewhere; (5) emigration of well-educated Filipinos (including teachers) to take jobs elsewhere; the economy of the Philippines is recovering

2. The Regional Hub, Singapore: Singapore began as an *entrepot* port city (a place where goods are imported, stored, and then shipped), but became the communications and financial hub of Southeast Asia; Singapore had a relatively mild recession in the 1990s, but its economy is healthy overall; Singapore's government is repressive and generally undemocratic

3. Malaysian's Insecure Boom: Malaysia is not as prosperous as Singapore, but has had rapid economic growth; the Malaysian economy is based on extraction of natural resources (hardwoods, palm oil and rubber, tin), and manufacturing, especially in labor-intensive high-tech sectors; Malaysia was hit hard by the Asian economic crisis of the late 90s; Malaysian leaders instituted currency controls, and they seem to be working; in Malaysia, there are geographical and ethnic disparities in economic opportunities; ethnic tensions (especially with minority Chinese) are an issue, and Malaysia has instituted affirmative action for *Bumiputra:* "sons of the soil," ethnic Malaysians

4. Thailand's Ups and Downs: Japanese firms helped fuel Thailand's boom in the 1980s; Thailand has a well-educated workforce, and it is politically stable, with a democratic government and a free press; many people did not benefit from the boom; the historical core (Bangkok) benefited most; Lao-speaking peoples in the northeast were left out; prostitution is a problem

5. Indonesian Economic Development: Indonesia's economy began to expand in the 1970s, due to oil exports and logging; Indonesia attracts foreign investment because of its low wages and abundant resources; the Indonesian government nurtured an indigenous,

technologically oriented business center; of all the countries of Southeast Asia, Indonesia was hardest hit by the 1990s recession; Indonesia's economic and social development are uneven, with Jakarta and Sumatra doing well, while elsewhere, poverty abounds

6. Divergent Economic Paths: Vietnam, Laos, and Cambodia (former French Indochina): this region at war from 1941–1975; Vietnam is the most prosperous of the three; Vietnam combines free market capitalism and communism, welcomes multinationals, encourages small entrepreneurs; Vietnam has seen an upsurge in economic activity; overall, Laos and Cambodia have low population densities and many natural resources; many people are still engaged in subsistence agriculture; Laos is building dams (for hydro power) and roads, but the repressive government is not attractive to many foreign investors; Cambodia's economy is becoming integrated with Thailand's, but it is not always popular with Cambodians; Cambodia is seeing a boom based on textiles, but elites have been taking land from peasants.

7. Burma's Troubled Economy ranks low in the region; Burma has abundant natural resources (oil, minerals, water, timber, fertile farmland), moderate population density, and reasonably well-educated people; Burma's economy has been stagnant since its independence (1948); Burma has a policy of self-contained "Buddhist Socialism;" today political instability and human rights violations deters investment in Burma

C. Globalization and the Southeast Asian Economy
   1. The economies of Southeast Asia are well integrated into the global economy
   2. Export-based economy makes this region vulnerable to world economic disruptions

D. Issues of Social Development
   1. Generally higher levels of economic development correlate with higher levels of social development
   2. In general, levels of social development in Southeast Asia are good, but Laos, Cambodia, and Burma are the exceptions
   3. Most governments in the region value education, so literacy rates are relatively high. University and technical education are not always readily available in Southeast Asia, forcing many to study abroad

VII. Conclusion
Southeast Asia is well integrated into the global economy. Although linguistic

153

and ethnic differences exist and sometimes cause conflict, the governments of this region have seen the value of unity, and have created ASEAN (The Association of Southeast Asian Nations).

PRACTICE MULTIPLE CHOICE QUIZ

1. Which phrase best describes the physical geography of Southeast Asia?
   a. Deserts and arid lands
   b. Extensive plains and grasslands
   c. Rolling hills and temperate forests
   d. Rugged terrain on the mainland and thousands of islands
   e. A combination of all of the above

2. In December 2004, which of the following environmental hazards caused tremendous loss of life and property damage in Southeast Asia?
   a. Volcanic eruption
   b. Typhoon
   c. Tsunami
   d. Forest fires
   e. Earthquake

3. What type of environmental damage is most prevalent in Southeast Asia?
   a. Deforestation
   b. Desertification
   c. Dessication of lakes
   d. Improper disposal of nuclear waste
   e. Salinization of farmland from irrigation

4. Which climate type dominates in Southeast Asia?
   a. Continental   b. Dry   c. Highland   d. Maritime   e. Tropical Humid

5. What types of agriculture are practiced in Southeast Asia?
   a. Swidden, growing subsistence crops
   b. Rice cultivation
   c. Plantation, growing rubber, cane sugar, pineapple, and copra
   d. Cash crops, opium
   e. All of the above

6.   All of the following statements about population in Southeast Asia are true, EXCEPT…
     a. The Malaysian government has instituted family planning programs similar to those found in Bangladesh
     b. The highest total fertility rates in the region are found in East Timor and Laos, and are explained by the low level of economic development
     c. Singapore and Thailand are the only countries in Southeast Asia with TFRs below replacement level
     d. In the Philippines, the Roman Catholic Church's stance against family planning contributes to a relatively high TFR
     e. Because of strong governmental support for family planning and improvements in education in Indonesia, it appears that this country will reach population stability before many other developing countries

7.   Which of the following groups is NOT one of the foreign influences on Southeast Asia?
     a. European colonists
     b. Immigrants from China
     c. Muslim merchants
     d. Russian conquerors
     e. South Asian immigrants

8.   Which country of Southeast Asia had its culture most profoundly chanced by European colonization?
     a. Vietnam, by the French
     b. Philippines, by the Spanish
     c. Malaysia, by Britain
     d. Indonesia, by the Netherlands
     e. Burma, by the British

9.   What attracted the earliest Europeans (the Portuguese) to Southeast Asia?
     a. Sugarcane   b. Rice   c. Nutmeg and cloves   d. Gold   e. Cotton

10. What European country became the dominant colonial power in Southeast Asia?
     a. Britain   b. France   c. Netherlands   d. Portugal   e. Spain

11. Which country of Southeast Asia was a colony of the United States from 1898 until 1946?
     a. Vietnam   b. Philippines   c. Laos   d. Cambodia   e. Indonesia

12. What was the name of the theory that held that if Vietnam fell to the communists, so would its neighbors, Laos and Cambodia, then Burma, Thailand, and perhaps all of Southeast Asia?
     a. Weakest link theory
     b. Manifest destiny
     c. Domino theory

d. Diffusion theory

e. Contagion theory

13. All of the following are examples of internal geopolitical tensions in Southeast Asia, EXCEPT...

    a. Rebelling ethnic groups in Burma seek to maintain their ethnic traditions and to overthrow a repressive regime

    b. Indigenous peoples of Papua resent what they see as Indonesian exploitation of their lands, and periodically rise up to demand independence

    c. In 2001, the U.S. sent a small number of counter-terrorism experts to the southern Philippines to help the Philippines battle Muslim separatists

    d. Growing tensions between the wealthy elite and sweatshop workers in Singapore have erupted in large-scale strikes and street protests

    e. Fundamentalist Muslims in the Aceh region of Indonesia have demanded the creation of a separate Islamic state

14. What Southeast Asian countries have had the greatest developmental successes in the region?

    a. Brunei and Indonesia

    b. Laos and Cambodia

    c. Philippines and Burma

    d. Singapore and Malaysia

    e. Vietnam and Thailand

15. What is the nature of the correlation between economic and social indicators in Southeast Asia?

    a. Countries with favorable economic indicators tend to have favorable social indicators

    b. Countries with favorable economic indicators have less favorable social indicators

    c. On the Mainland, favorable economic indicators are correlated with favorable social indicators, but throughout Insular Southeast Asia, favorable economic indicators are correlated with unfavorable social indicators

    d. Throughout Insular Southeast Asia, favorable economic indicators are correlated with favorable social indicators, while on Mainland Southeast Asia, favorable economic indicators are correlated with unfavorable economic indicators

    e. There is no correlation between social and economic indicators

Answers: 1-D; 2-C; 3-A; 4-E; 5-E; 6-A; 7-D; 8-B; 9-C; 10-C; 11-B; 12-C; 13-D; 14-D; 15-A

Chapter 14
AUSTRALIA AND OCEANIA

LEARNING OBJECTIVES
- This chapter covers Australia and Oceania, which also includes New Zealand and a sweeping collection of islands that reaches halfway into the Pacific Ocean.
- The student should understand the unique geography of archipelagos (island groups) and the equally unique cultural adaptations that the residents of this region have implemented.
- The student should understand the relationships between the indigenous peoples of this far-flung region and the European peoples who have come to dominate much of this region and be able to compare this situation to what is found in the United States.
- In addition, the student should understand the following concepts and models:

  · Atoll
  · Archipelago
  · Coral reefs and atoll islands
  · Tsunami
  · Exotics and extinction
  · Aborigines and Maoris
  · Pidgin English

CHAPTER OUTLINE
I. Introduction
  A. Oceania contains two distinctly different worlds. Australia and New Zealand are culturally and economically linked with Europe, even though the landforms are distinctly not European. The rest of the region (Oceania) consists of island chains covering the South Pacific. This region is subdivided into three regions: Polynesia, Melanesia, and Micronesia. Oceania is united by historical isolation, culture clashes, and a relatively new political geography.
  B. Australia and New Zealand dominate this region; Australia has a huge, dry interior (the outback) that is thinly settled. New Zealand has mountains that limit settlement there.
  C. There are three main archipelagos (island groups) in Oceania: Melanesia (dark islands) is culturally complex; Polynesia (many islands) is linguistically unified; Micronesia (small islands) includes microstates and Guam.

II. Environmental Geography: A Varied Natural and Human Habitat
  A. Australian Environments
    1. Regional landforms (Fig. 14.1), major landforms in Australia: (1) Western Plateau covers more than half the country and geologically represents the remnants of a shield formation that once connected to

Antarctica; (2) the Interior Lowland Basins stretch north to south for 1,000 miles; most of the region is a flat, featureless plain with dry lake beds and by stream valleys where water is rare; (3) Eastern Highlands are located on the Pacific Rim coast and include the narrow, highly settled coastal plain; (4) Great Barrier Reef is underwater, off the coast of Queensland; it is the one of the world's most remarkable examples of coral reef building

2. Climate and Vegetation: the north has a monsoon climate with dry winters and wet summers, producing tropical woodlands, thorn forests mixed with open grasslands; Central Australia is dry, with little rainfall (less than 1 inch per year), producing scrub vegetation; Southeastern Australia has year-round rainfall averaging 40–60 inches per year producing forests along the coast; Southwestern Australia Mediterranean climate, with dry winters and wetter summers, producing mallee-eucalyptus woodland with little economic value

3. An unusual zoogeography: isolation and genetics created a mammal group in Australia based on marsupials; bird life is highly varied

B. New Zealand's Varied Landscape
  1. Landforms include numerous active volcanoes and geothermal features; and the Southern Alps
  2. Zoogeography: like Australia, New Zealand's isolation produced unique flora and fauna; 85 percent of New Zealand's native trees and seed plants are found nowhere else on earth; bats are the only native mammal
  3. North Island Environments: mainly subtropical, microclimatic variations are found on the volcanic peaks; most of the flora has been replaced by introduced European species
  4. South Island Environments: are typical mid-latitude at the north end, while the southern end experiences a winter chill; the west side of the Southern Alps is tropical rain forest, while the east side is grassland

C. The Oceanic Realm
  1. Creating Island Landforms: Melanesia and New Zealand formed from continental rock; most of the islands of Polynesia and Micronesia were formed by volcanic activity on the ocean floor with no connection to larger landmasses; Hawaii is *a high island* (formed by larger active and recently active volcanoes); the Hawaiian archipelago is also an example of a *geological hot spot,* where moving oceanic crust passes over a supply of magma, creating a chain of volcanic uplifts; *low islands* are formed out of the eroded coral reef; when these islands form a ring around a shallow central body of water, they form an *atoll*
  2. Island Climates: some islands have high rainfall and dense tropical forests; low-lying atolls receive much less rain than the high islands, and sometimes experience water shortages; high islands have lush tropical forests

D. Environments at Risk: this region faces challenges that include seismic hazards, periodic Australian droughts, and tropical cyclones

1. Global Resource Pressures: mining has had a negative impact on many parts of Australia and Oceania, where semiarid regions are susceptible to metals pollution; deforestation has caused the loss of vast stretches of eucalyptus woodlands to create pastures in Australia; elsewhere (e.g., Papua New Guinea), logging pressures cause deforestation

2. Nonnative Plants and Animals: exotic (non-native) animals and plants have been introduced to the region, and they have had a detrimental effect on native animals; in Australia, where the environment lacks the diseases and predators that keep rabbits in check elsewhere, rabbits have reached plague proportions, as large pieces of land were stripped of vegetation by the bunnies; sheep and goats have accelerated soil erosion and desertification; island environments have also experienced problems after the introduction of exotic animals; for example, the Moa (a bird larger than the ostrich) became extinct after Polynesian settlers to New Zealand hunted the birds, burned their habitat, and brought (accidental) rats that ate the eggs of the Moa; during the second wave of migration, European settlers brought non-native species that often competed successfully with native species

3. Global Warming in Oceania: some researchers predict that global warming may cause higher global temperatures to melt polar ice caps, which will in turn raise ocean levels and drown many islands in Oceania; some islands at low elevations in this region are already experiencing slight increases in water level and increased coastal erosion; Australia ratified the Kyoto Protocol in 2007, leaving the U.S. as the only industrialized country that has not signed it.

III. **Population and Settlement: A Diverse Cultural Landscape**

A. Contemporary Population Patterns (Fig. 14.17)

1. Australia: has 21 million people, it is highly urbanized, most people live in the subtropical south and east; Aborigines (the indigenous people) live in the arid center

2. New Zealand has 4.2 million people, and 70 percent of them live on the North Island

3. Oceania is overwhelmingly rural; the largest city is Honolulu, and its population is a result of migration from the American mainland

B. Historical Settlement (Fig. 14.18)

1. Peopling the Pacific: 40,000 years ago, the ancestors of the *Aborigines* (native Australians) came to the region by boat; Melanesia was settled 3,500 years ago by people who had perfected long-distance sailing and navigation

2. European Colonization: European explorers "discovered" this region in 15th century; *Australia* was established as a prison colony in 1788 and the

British government, who expelled Aboriginal peoples from the land, supported further migration; *New Zealand* was settled by whalers and sealers, and Britain settled the region beginning in 1840; tensions between Maoris (native new Zealanders), and British settlers led to wars from 1845–1870; initially, a powerful *Hawaiian* ruling family prevented Euro-American claims to the islands, in 1898, Hawaii became a U.S. territory; other parts of Oceania have had little European settlement

C. Modern Settlement Landscapes
   1. The Urban Transformation: urbanization began in Australia and New Zealand in the 20th century; most urban areas have vibrant downtowns with low crime rates; every major urban area in the region includes coastal features, waterfront districts, and harbors; urbanization elsewhere in Oceania is different; often there is a lack of housing, street crime is prevalent, and alcoholism is a problem; urban areas are growing rapidly through immigration from nearby rural areas and islands
   2. The Rural Scene: in Australia, sheep and cattle ranching are significant; some sugar cane and truck farming occurs near Perth and in Murray Valley; viticulture (grape cultivation) is increasing; in New Zealand, sheep ranching and dairying are important; in Oceania, subsistence farming of taro, sweet potatoes, coconuts, and bananas occurs; cash crops such as coffee, cocoa, and sugar cane are also important

D. Diverse Demographic Paths
   1. Australia and New Zealand had high population growth in the early part of the 20th century; today, both countries have low birthrates; retirement communities and suburbs are growing, and older industrial areas are in decline
   2. Oceania: population growth exceeds 2 percent per year; small islands tend to have high population densities, compounded by migration to urban areas

IV. Cultural Coherence and Diversity: A Global Crossroads
   A. Multicultural Australia
      1. Aboriginal Imprints: currently, about 2 percent of Australia's population are Aborigines, who followed a hunting-gathering way of life for thousands of years; Europeans pushed Aborigines into the arid central region of Australia; many Aborigines are employed in urban areas, Christianity is their major religion, and only 13 percent speak their native language (Fig. 14.27); Australia has programs in place to preserve Aboriginal culture
      2. A land of Immigrants: 70 percent of Australians are of Irish and British descent; *Kanakas* (laborers from islands in Oceania such as the Solomons and New Hebrides) were imported to work on farms; the "White Australia" policy limited immigration into Australia to Europe and North America until 1973; most current

immigrants to Australia come from Asia; Australia's One Nation Party wants to restrict immigration into Australia

B. Cultural Patterns in New Zealand
1. European culture is dominant in the country
2. Maori make up 8 percent of the country, are found mainly on North Island, and are committed to preserving their culture; Maori is an official language of New Zealand
3. Asian immigrants now make up of 5 percent of New Zealand's population
4. Movies made in New Zealand include the *Lord of the Rings* trilogy and *The Piano*

C. The Mosaic of Pacific Cultures
1. Language Geography: most languages in the region are Austronesian; Papua New Guinea has 1,000 different languages and holds some of the few remaining *uncontacted peoples* (cultural groups that have yet to be "discovered" by the Western world)
2. Village Life: settlements (villages) in Melanesia usually have fewer than 500 people, and life there revolves around farming; Polynesia has class-based societies; violent warfare was common before the arrival of Europeans
3. External Cultural Influences: Europeans, Americans, and Asians influenced Hawaii, Guam, and Fiji; local languages are being supplanted by *Pidgin English*, languages formed from local languages and English; indigenous religions have been replaced by the Christian religions of settlers; in Melanesia, mergers of Christianity and animist religions have merged in the *cargo cults* (quasi-animist religions of Melanesia that originated with military cargo supply dumps during World War II); tourism is a source of revenue for many Pacific islands, bringing more contact with outsiders

V. **Geopolitical Framework: A Land of Fluid Boundaries**
A. Geopolitics in this region reflect a complex interplay of local, colonial-era, and global-scale forces
B. Shifting Geopolitical Patterns
1. Indigenous Patterns: Aboriginal Australia was organized around fluid groups of 30–60 related people; hunting gathering was important. In Melanesia, the groups were larger, sedentary groups based on kinship. Polynesia and Micronesia had highly structured chiefdoms, centralized kingdoms, with one person in complete control; this also occurred in Hawaii
2. An Imposed Colonial Framework: Australia and New Zealand were settled by immigrants from Europe and became British colonies. Oceania was divided up among different global powers in the 19th century, when their values as refueling centers and telegraph cable relay points became evident. The countries that divided Oceania were France (Tahiti, Polynesia), Germany (northeastern New Guinea, parts of Micronesia, Solomon Islands, and Samoa), Great Britain

(western Polynesia, Fiji, southern Solomon Islands), United States, (Hawaii, Guam).

3. Roads to Independence
Australia and New Zealand became independent in 1901 and 1907 and are members of British Commonwealth; both countries are considering becoming republics
Japan, France, and the United States have all controlled territory in the region
In *Oceania,* former colonies gained their independence in the 1970s
New Zealand and France still control territory in the region

C. Persisting Geopolitical Tensions
1. Native Rights in Australia and New Zealand: Australia established Aboriginal Reserves in Central Australia; passed the Native Title Bill, which paid Aborigines for land taken from them and allows them to gain title to unclaimed land and to deal with mining companies; New Zealand's Maori claim land rights to much of North and South Islands and want name of country changed to Aotearoa (Maori name that means "Land of the Long White Cloud"); there have been conflicts between the Maori and the government on this issue
2. Conflicts in Oceania: Fiji military leaders overthrew a government controlled by South Asians in 1987; the current constitution favors the native Fijians; in Papua New Guinea, local conflicts occur between different culture groups; natives of the island are concerned that

government is exploiting abundant natural resources on the island; in French Polynesia, an independence movement is underway; in the Tuamotu Archipelago, local protests have opposed nuclear testing in the atolls

D. A Regional and Global Identity?
1. Australia and New Zealand are the political leaders in the Pacific; both countries mediate conflicts within the region and maintain links with both North America and Southeast Asia
2. France and the United States. maintain close links with current and former colonies

VI. Economic and Social Development: A Difficult Path to Paradise
A. Uncertain Avenues to Affluence
1. The Australian Economy: Australia's historical affluence was dependent on export of raw materials (copper, iron ore, bauxite, nickel, gold, lead, and zinc); Australia has little manufacturing and high technology industry; tourism is becoming a growth industry; wealth is unevenly distributed (concentrated in major cities, and higher poverty levels among Aborigines; their incomes are only 65 percent of the national average)
2. New Zealand's Economic Challenge: New Zealand relies on traditional agricultural exports for revenues; in the 1990s, its economy stagnated, and it adopted drastic economic reforms; today it is one of the world's most market-oriented

countries; it is not clear how successful this strategy will be
3. Oceania's Economic Diversity: Melanesia is the least developed and poorest region of Oceania; most countries are dependent on exports of coffee, sugarcane, and coconut; in Micronesia mining is important, and others support themselves with subsistence agriculture; in Polynesia, some of the countries receive subsidies from France and the United States; in Hawaii, French Polynesia, and Guam, tourism is important
B. The Global Economic Setting
1. Oceania has reduced economic ties with North America and Asia, while Australia and New Zealand are members of the Asia-Pacific Economic Cooperation (APEC) Group— organized to form economic ties throughout the region
C. Enduring Social Challenges
1. Australia and New Zealand's people are susceptible to most of the typical problems of the industrialized world; cancer and heart disease are the leading

causes of death, and alcoholism is a persistent problem
2. Aborigines and Maori have many more problems; schooling is irregular for many natives, and discrimination against native populations is a continuing problem

VII.   Conclusion

Due to their location and population, Australia and Oceania are in a peripheral position in world affairs. The region is attempting to globalize, with growing attachments to Asia. Australia is the dominant country of the region and is the financial and industrial center of the South Pacific. However, it has an economic output equal to Texas and has suffered through a number of recessions. New Zealand is working toward closer ties with both Australia and Asia, while coming to terms with native Maori and Asian immigrants. Oceania is a crossroads of Pacific, Asian, and European cultures. This region has resources through their territorial rights over oceans and investment from foreign countries. However, they do not have much global political power, with former colonial powers exerting much power sometimes unwelcome.

PRACTICE MULTIPLE CHOICE QUIZ

1. Why do seismic hazards, periodic droughts in Australia and violent tropical cyclones pose a greater threat now than they did in the past?
   a. Because of global warming, these events are becoming more frequent
   b. Shifting tectonic plates are causing the seismic events to become stronger, while global warming is causing the tropical cyclones to become more powerful and the droughts to last longer
   c. New settlements have made increasing populations vulnerable to these problems
   d. A and B above
   e. A, B, and C above

2. Which of the following is NOT a threat to the environment of Australia/Oceania?
   a. Subsidence
   b. Fallout from nuclear testing
   c. Major mining operations
   d. Logging by transnational corporations
   e. Clear-cutting forests to create pastures

3. Which of the following exotic species were introduced to Australia/Oceania, either intentionally or accidentally, and caused significant problems in the region?
   a. Brown tree snake
   b. Rabbit
   c. Rat
   d. Sheep
   e. All of the above

4. What is the driest part of Australia/Oceania
   a. Australia's center
   b. Fiji
   c. New Zealand's North Island
   d. Tasmania
   e. The western coastline of Vanuatu

5. Volcanic eruption is the first step in the creation of which of the following features?
   a. Atoll
   b. High islands
   c. Tsunamis
   d. A and B above
   e. A, B, and C above

6. Where do the majority of the people of the region of Australia/Oceania live?
   a. Australia
   b. Hawaii
   c. New Zealand
   d. Marshall Islands
   e. Rural villages throughout the region

7. All of the following statements about Aborigines are correct, EXCEPT…
   a. Aborigines originated in Southeast Asia
   b. Aborigines were foragers in and hunters in Australia when the first European settlers arrived
   c. In Tasmania, Aborigines were hunted down and killed
   d. Aborigines were run off their lands, and although their populations dwindled, they survived in the Australian Outback
   e. To this day, most Aborigines live in the Outback

8. Which of the following are NOT characteristics of cities in Australia/Oceania?
    a. Australia and New Zealand's cities have many slums and high crime rates, just as in North America
    b. In countries other than Australia and New Zealand, rapid growth has produced many of the usual challenges associated with underdevelopment
    c. Major cities in Australia and New Zealand center on downtowns that often resemble those of metropolitan North America
    d. Much of the urban transformation in the region occurred during the Twentieth Century
    e. The urban landscape of Port Moseby in Papua New Guinea has inadequate housing and infrastructure

9. Which of the pairings of agricultural activity and country is INCORRECT?
    a. Copra, cocoa, and coffee – Solomon Islands
    b. No agriculture (too dry) – Australia
    c. Sheep – New Zealand
    d. Shifting cultivation – New Zealand
    e. Sugarcane plantations – Hawaii

10. All of the following statements about the Maori are correct, EXCEPT...
    a. After declining with initial European contacts and conflicts, the Maori population began to rebound in the 20th Century
    b. Maori are trying to assimilate into the European populations of New Zealand
    c. The Maori are most numerous on New Zealand's North Island
    d. The Maori are an indigenous minority group in New Zealand
    e. The Maori are more numerically important and culturally visible in New Zealand than the Aboriginal counterparts in Australia

11. What is the major difference between the boundaries created by indigenous peoples of Australia/Oceania and those created by European colonists?
    a. Indigenous people naturally evolved clear ethnographic boundaries; Europeans replaced them with physiographic boundaries
    b. Indigenous peoples had established firm boundaries with their neighbors; Europeans eliminated these boundaries
    c. Indigenous peoples preferred nuanced, often fluid boundaries; Europeans preferred more precise, yet unstable borders
    d. Indigenous peoples preferred physiographic boundaries; Europeans preferred geometric boundaries
    e. Indigenous peoples used local landmarks to mark boundaries; Europeans used fences

12. Which foreign country continues to maintain a colonial presence in Australia/Oceania, causing tensions in New Caledonia and resuming nuclear testing in the Tuamotu Archipelago?
    a. United States   b. Portugal   c. Netherlands   d. France   e. Britain

13. What is Aoteroa?
   a. The name of the official airline of Australia
   b. The name that the Maoris call New Zealand
   c. The indigenous peoples of Tasmania
   d. An indigenous, region-wide environmental movement in Australia/New Zealand
   e. A flightless bird with hairy feathers native to Australia

14. Which country of Australia/Oceania is grouped among the world's developed countries?
   a. Australia
   b. New Zealand
   c. French Polynesia
   d. A and B above
   e. A and C above

15. Which of the following pairings of country and economic activity is incorrect?
   a. Solomon Islands – fish canning, coconut processing
   b. New Caledonia – nickel mining
   c. Nauru – phosphate mining
   d. Hawaii – assembly plant industrialization
   e. Fiji – sugar production

Answers: 1-C; 2-A; 3-E; 4-A; 5-D; 6-A; 7-E; 8-A; 9-B; 10-B; 11-C; 12-D; 13-B; 14-D; 15-D